Smart Zero-energy Buildings and Communities for Smart Grids

Engineering, Energy and Architecture Set

coordinated by
Lazaros E Mavromatidis

Volume 9

Smart Zero-energy Buildings and Communities for Smart Grids

Edited by

Nikos Kampelis
Denia Kolokotsa

ISTE

WILEY

First published 2022 in Great Britain and the United States by ISTE Ltd and John Wiley & Sons, Inc.

ISTE Ltd
27-37 St George's Road
London SW19 4EU
UK

www.iste.co.uk

John Wiley & Sons, Inc.
111 River Street
Hoboken, NJ 07030
USA

www.wiley.com

Library of Congress Control Number: 2021948475

British Library Cataloguing-in-Publication Data
A CIP record for this book is available from the British Library
ISBN 978-1-78630-684-5

MIX
Paper from responsible sources
FSC www.fsc.org FSC® C013604

Contents

Preface xi
Nikos KAMPELIS

List of Acronyms xv
Nikos KAMPELIS

Chapter 1. The Role of Smart Grids in the Building Sector 1
Denia KOLOKOTSA

1.1. Smart and zero-energy buildings 2
1.1.1. Smart metering 3
1.1.2. Demand response (DR) 4
1.1.3. Distributed systems. 6
1.2. Smart and zero-energy communities 6
1.3. Conclusion and future prospects 10

Chapter 2. Integrated Design (ID) Towards Smart Zero-energy Buildings and Smart Grids 13
Theoni KARLESSI, Pietro MURATORE, Luca VENEZIA, Laura STANDARDI, Klemens LEUTGÖB and Anne Sigrid NORDBY

2.1. Introduction. 15
2.2. Methodology 16
2.3. Integrated design in smart and zero-energy buildings 17
2.4. ID process principles and guidelines 19
2.4.1. Benefits. 22
2.4.2. Barriers 23

2.5. Scope of services 24
2.6. Remuneration models. 26
2.7. Application of evaluation tools 28
2.8. Sustainability certification 29
2.9. Consultancy and quality assurance 30
2.10. Measurement of design quality criteria. 31
2.11. Defining a client's objectives 33
2.11.1. Capital cost reduction. 34
2.11.2. Delivery risk reduction. 35
2.12. Defining the tenant's objectives 35
2.12.1. Operational cost reduction. 36
2.12.2. Building unsuitability risk reduction. 36
2.13. Best practice sites 37
2.13.1. Alexandros N. Tombazis and Associates Architects S.A. office building 37
2.13.2. APIVITA Commercial and Industrial S.A. 42
2.13.3. Stavros Niarchos Foundation Cultural Center 46
2.13.4. Karelas Office Park 50

Chapter 3. Data Analysis and Energy Modeling in Smart and Zero-energy Buildings and Communities 55

Nikos KAMPELIS, Konstantinos GOBAKIS, Vagias VAGIAS, Denia KOLOKOTSA, Laura STANDARDI, Daniela ISIDORI, Cristina CRISTALLI, Fabio Maria MONTAGNINO, Filippo PAREDES, Pietro MURATORE, Luca VENEZIA, Marina Kyprianou DRACOU, Alaric MONTENON, Andri PYRGOU, Theoni KARLESSI and Mat SANTAMOURIS

3.1. Energy signature for the NTL of Cyprus Institute 55
3.2. Athalassa Campus and the NTL building 57
3.2.1. Methodology 61
3.2.2. Description of the Novel Technology case study 63
3.2.3. Data exploration 68
3.2.4. Correlation matrix 71
3.2.5. Regression model. 72
3.3. Linear Fresnel solar collector at the NTL building, Cyprus Institute . . 85
3.3.1. Development of the NTL model 90
3.3.2. Energy performance analysis in the NTL. 92
3.3.3. Discussion 100
3.4. Conclusion 101

Chapter 4. On the Comparison of Occupancy in Relation to Energy Consumption and Indoor Environmental Quality: A Case Study 103
Margarita Niki ASSIMAKOPOULOS, Nikolaos BARMPARESOS, Alexandros PANTAZARAS, Theoni KARLESSI and Siew Eang LEE

4.1. Introduction 103
4.2. Methodology 104
4.3. Description of the case building 105
4.4. Description of the experimental procedure 105
4.5. Results 106
4.5.1. Investigation of energy consumption and indoor air quality 106
4.5.2. Days of special interest – high occupancy 110
4.5.3. Days of special interest – increased energy consumption 112
4.6. Discussion and concluding remarks 112

Chapter 5. Indoor Environmental Quality and Energy Consumption Assessment and ANN Predictions for an Integrated Internet-based Energy Management System Towards a Zero-energy Building 115
Denia KOLOKOTSA

5.1. Introduction 115
5.2. Description of the SDE buildings 116
5.2.1. General information 116
5.2.2. Monitoring activities for SDE 3 118
5.3. The power loads and hourly energy consumption 118
5.4. Indoor environmental quality 118
5.4.1. Thermal comfort assessment – time series analysis 127
5.4.2. Indoor air quality 129
5.4.3. The indoor illuminance levels 129
5.5. Cross correlation 135
5.6. Prediction using artificial neural networks (ANN) 136
5.6.1. Prediction of outdoor temperature 137
5.6.2. Prediction of relative humidity 138
5.6.3. Prediction of power loads 139
5.7. Specifications for an integrated internet-based energy management system toward a zero-energy building 141
5.7.1. The phases of the internet-based energy management system for SDE 142
5.7.2. Integration of software and prediction algorithms 149
5.8. Conclusion 149

Chapter 6. Objective and Subjective Evaluation of Thermal Comfort in the Loccioni Leaf Lab, Italy 151
Marina LASKARI, Francesco CARDUCCI, Daniela ISIDORI, Martina SENZACQUA, Laura STANDARDI and Cristina CRISTALLI

6.1. Introduction 151
6.2. Background information 152
6.3. Methodology 153
6.3.1. Subjective measurements 154
6.3.2. Objective measurements 154
6.3.3. Combined analysis of objective and subjective measurements . . . 155
6.3.4. User preferences and satisfaction with internal conditions 157
6.4. Collection of building background data 157
6.5. Collection of monitored data 160
6.6. Right-Now questionnaire survey 162
6.7. Results 166
6.7.1. Analysis of MyLeaf measurements 167
6.7.2. Analysis of Comfort Meter measurements 173
6.7.3. Analysis of Right-Now survey responses 176
6.7.4. Respondent characteristics and thermal comfort 184
6.7.5. Combined analysis of objective and subjective measurements . . . 187
6.7.6. Correlation analysis for MyLeaf and Right-Now survey measurements 190
6.7.7. Correlation analysis for objective and subjective measurements (Research for Innovation office space) 191
6.7.8. Comparison between objective and subjective thermal sensation measurements 195
6.7.9. Determination of acceptable and unacceptable conditions 196
6.8. Conclusion 197

Chapter 7. Smart Meters and User Engagement in the Leaf House 199
Niki GAITANI

7.1. Introduction 199
7.2. Methodology 200
7.3. Analysis of user engagement 201
7.3.1. Development of the questionnaire 201
7.3.2. Leaf House case study 203
7.4. Results 210
7.4.1. Demographics, socioeconomics 210
7.4.2. Physiological, social and behavioral aspects 212

7.4.3. Information level 214
7.4.4. Health and comfort 215
7.4.5. Living situation 217
7.5. Conclusion 218

Chapter 8. Integration of Energy Storage in Smart Communities and Smart Grids 221
Denia KOLOKOTSA, Nikos KAMPELIS, Angeliki MAVRIGIANNAKI, Marco GENTILOZZI, Filippo PAREDES, Fabio Maria MONTAGNINO and Luca VENEZIA

8.1. Energy storage systems in smart grids 223
8.1.1. Electrical and electrochemical energy storage in smart grids 223
8.1.2. Mechanical energy storage in smart grids 228
8.1.3. Thermal energy storage in smart grids 231
8.2. Energy storage and smart grids: case studies 234
8.2.1. Case study 1: the Leaf Community smart grid energy storage system 234
8.2.2. Case study 2: energy storage of CSP and integration with smart grids. 244
8.3. Conclusion and future prospects 261

Conclusion and Recommendations 263
Nikos KAMPELIS

References 267

List of Authors 283

Index 287

Preface

Smart and zero-energy buildings and communities have a major role to play in the evolution of the building sector and of the electric grid (i.e. the smart grid) toward the necessary transition in line with current carbon neutrality policies, climate change mitigation and adaptation. In this sense, the goal for the reduction of greenhouse emissions in the built environment can be pursued through a holistic approach whereas the consumption of energy in buildings is drastically reduced. State of the art in this field relates, on the one hand, to the implementation of energy efficient design methodologies and innovative technologies which guarantee the maximum energy saving potential in buildings. On the other hand, the advancement of ICT technologies, along with the integration of renewables and storage at building and at district level, provide the means for zero or positive net energy buildings and districts by producing, storing, managing and exchanging energy at the local level. In this direction, the challenges related with the volatility of renewable energy sources at grid level can become more manageable. New and emerging roles and services linking the building sector with smart grids in the future should be transparent and promote sustainability. This requires inspiring, fair, effective and innovative policies providing the roadmap for this transition and major research, innovation and training initiatives that will (1) support the market in providing solutions supporting mass scale deployment of environmentally friendly, energy efficient technologies and (2) educate the society about the importance of this transition and the role each and every one of us has to play.

Why this book?

This book is a collaborative work between research and industrial partners in the framework of the Smart GEMS Marie Curie project (2015–2020). Research and training activities in Smart GEMS were implemented through the physical mobility (secondments) of staff between academic and industrial partners during a five-year period. It provided academic staff with the opportunity to expand their views by working with leading companies in the fields of advanced building/district energy management, renewable technologies and storage integration, as well as in the production of smart metering systems. Likewise, industrial personnel had the opportunity to work in different academic environments across Europe and establish a more coherent understanding about the broader scientific and technological capabilities outside their field of specialization, thus complementing their field of expertise and in various cases diving deeper into co-developing and exploring state-of-the-art techniques and methodologies. This book provides a thorough reading of many of the concepts dealt with in Smart GEMS through the close collaboration of industrial and academic partners. Adding to the first publication, entitled *Smart Buildings, Smart Communities and Demand Response*, published by ISTE Ltd and John Wiley & Sons, this book provides complementary material and an insight into the fields of smart grids, smart and zero-energy buildings, integrated design, data analysis and energy modeling, indoor environmental quality, user engagement and energy storage in smart communities and smart grids.

Who is this book for?

This book focuses on smart nearly zero-energy buildings (NZEB), smart communities and smart grids. Therefore, it is mainly valuable for experts, professionals and researchers with an interest in (1) energy efficient buildings and communities; (2) smart building systems and innovative applications; and (3) integration of renewable energy technologies and storage at the building and district level.

Structure

First, smart grids are defined and their role in integrating more renewable energy sources, smart buildings and distributed generators, thereby revolutionizing the electric power network, is presented. Concepts such as smart and zero-energy buildings and communities along with smart metering, demand response and distributed systems are outlined to provide the state of the art, opportunities and the challenges for minimizing buildings' carbon footprint.

Second, the main principles of integrated design and decision making for smart zero-energy buildings and grids are documented and explored with the aid of best practice examples. Benefits, barriers and methodologies for addressing the potential and evaluating the impact of integrated design are explained.

In Chapter 3, with the aid of case studies, we present data analysis and energy modeling of smart and zero-energy buildings and communities for the evaluation and management of energy operations in buildings integrating innovative renewable energy technologies.

In Chapter 4, the impact of human presence to the energy consumption and indoor air quality within an educational building of the National University of Singapore (NUS) is investigated. An experimental campaign took place and data was analyzed with respect to the energy consumption and air quality of three rooms, each one with different usage and occupancy. The impact of different occupancy patterns on the energy demand, the illuminance of the building, as well as the internal levels of temperature, relative humidity and CO_2, are examined.

In Chapter 5, the energy consumption and indoor environmental quality of one of the three buildings of the Design and Environment School (SDE3) of the National University of Singapore is evaluated and cross-correlated based on a different perspective. Prediction algorithms based on artificial neural network models are tested.

In Chapter 6, objective and subjective evaluation of thermal comfort is analyzed in the context of a unique smart zero-energy industrial facility in Italy.

In Chapter 7, the user engagement of residents in a smart zero-energy building in the same area of Italy is investigated in order to provide the framework for analyzing individual preferences, identifying consumption patterns and assessing the utilisability of information provided to users as well as how effective this is in supporting behavioral change.

Chapter 8 deals with the integration of energy storage in smart communities and smart grids. The various energy storage technologies are presented including electrical, mechanical, chemical and thermal. Energy storage and optimization of its utilization in smart grids integrating renewable energy technologies is explored through state-of-the-art case studies.

Finally, the conclusion outlines the main and overall conclusions and recommendations stemming from the findings of the presented research.

Acknowledgments

The editors express their deepest appreciation and gratitude to all partners, personnel, and researchers for their unique contributions, time and efforts which altogether resulted in making this publication happen. We are also very thankful to the European Commission and the EU taxpayer for devoting the necessary financial resources for the implementation of the Smart GEMS project.

Nikos KAMPELIS
November 2021

List of Acronyms

AMI	Advanced Metering Infrastructure
ANN	Artificial Neural Network
AP	Accredited Professional
ATES	Aquifer Thermal Energy Storage
BAS	Building Automation System
BCVTB	Building Controls Virtual Test Bed
BEMS	Building Energy Management System
BIM	Building Information Modeling
biPV	building-integrated PhotoVoltaics
BMS	Building Management System
CAES	Compressed Air Energy Storage
CDD	Cooling Degree Days

Chapter written by Nikos KAMPELIS.

CHP	Cogeneration of Heat and Power
COP	Coefficient Of Performance
CPC	Compound Parabolic Collector
CSP	Curtailment Service Provider
Cv	Coefficient of variance
DA	Day-Ahead
DC	Direct Current
DER	Distributed Energy Resources
DG	Diesel Generator
DHW	Domestic Hot Water
DNI	Direct Normal Irradiance
DR	Demand Response
DSG	Direct Steam Generation
DSM	Demand Side Management
EED	Energy Efficiency Directive
EER	Energy Efficiency Ratio
EES	Electrical Energy Storage
EMS	Energy Management System
EPBD	Energy Performance Buildings Directive
ES	Energy Signature

ESEER	European Seasonal Energy Efficiency Ratio
ETL	Extract, Transform, Load
EV	Electric Vehicles
FC	Fuel Cell
FCU	Fan Coil Units
FMU	Functional Mock-up Units
G2V	Grid-to-Vehicle
GA	Genetic Algorithm
GSHP	Ground Source Heat Pumps
HDD	Heating Degree Days
HESS	Hybrid Energy Storage Systems
HMI	Human Machine Interface
HPS	Hydro-Pumped Systems
HRU	Heat Recovery Units
HTF	Heat Transfer Fluid
HVAC	Heating, Ventilation, Air Conditioning
HVDC	High Voltage Direct Current
IAM	Incident Angle Modifier
ICT	Information and Computer Technology
ID	Integrated Design

IED	Integrated Energy Design
IoT	Internet of Things
k-NN	k-Nearest Neighbor
KPI	Key Performance Indicator
LCA	LifeCycle Analysis
LCC	LifeCycle Cost
LCCA	LifeCycle Cost Assessment
LFC	Linear Fresnel Collectors
LFR	Linear Fresnel Reflector
LOLP	Loss Of Load Probability
MAPE	Mean Average Percentage Error
MS	Member States
MS	Molten Salt
MS-TES	Molten Salt Thermal Energy Storage
NARX	Nonlinear AutoRegressive network with eXogenous input
NUS	National University of Singapore
NZEB	Nearly Zero-Energy Building
ORC	Organic Rankine Cycle
PCM	Phase Change Material
PEV	Plug-in Electric Vehicles

PLC	Programmable Logic Controller
PMP	Performance Measurement Protocols
PMV	Predicted Mean Vote
POD	Point Of Delivery
PPD	Percentage of People Dissatisfied
PSO	Particle Swarm Optimization
PTC	Parabolic Trough Collectors
PV	PhotoVoltaic
R	Pearson's coefficient
RES	Renewable Energy Sources
RforI	Research for Innovation
RH	Relative Humidity
RMSE	Root Mean Squared Error
SCTF	Single Coil Twin Fan
SDE	School of Design and Environment
SEER	Seasonal Energy Efficiency Ratio
SHC	Solar Heating and Cooling
SME	Small and Medium Enterprises
SMERC	SMart grid Energy Research Center
SMES	Superconducting Magnetic Energy Storage

SPSS	Statistical Package for Social Sciences
TES	Thermal Energy Storage
V2B	Vehicle-to-Building
V2G	Vehicle-to-Grid
VRFB	Vanadium Redox Flow Batteries
WT	Wind Turbine
ZEB	Zero-Energy Buildings

1

The Role of Smart Grids in the Building Sector

A smart grid is a dynamically interactive real-time infrastructure concept that encompasses the many visions of the stakeholders of diverse energy systems (El-Hawary 2014). Smart grids are electrical power grids that are more efficient and more resilient, and therefore "smarter", than existing conventional power grids. The smartness is focused not only on the elimination of blackouts, but also on making the grid greener, more efficient, adaptable to customers' needs, and therefore, less costly (El-Hawary 2014; Giordano *et al.* 2013). Smart grids incorporate innovative IT technology that allows for two-way communication between the utility and its customers/users. As a result, the sensing along the transmission lines and the sensing from the customer's side is what makes the grid "smart".

Like the Internet, the smart grid consists of controls, computers, automation, new technologies, smart buildings and equipment working together, but in this case these technologies will work with the electrical grid to respond digitally to the quickly changing energy demands of the users. Therefore smart grids create an exceptional opportunity for the support of the development of smart zero-energy buildings and communities, and they offer a step towards the Internet of Things (IoT) for the Energy and Building Industry (Chen *et al.* 2013; Zhen *et al.* 2012).

Chapter written by Denia KOLOKOTSA. This chapter is based on a previous work by the author, "The role of smart grids in the building sector", *Energy and Buildings*, 116, 703–708, 2016, reproduced with permission from Elsevier.

For a color version of all figures in this chapter, see www.iste.co.uk/kolokotsa/smartbuildings2.zip.

Smart grids open the door to new applications with far-reaching interdisciplinary impacts: providing the capacity to safely integrate more renewable energy sources (RES), smart buildings and distributed generators into the network; delivering power more efficiently and reliably through demand response and comprehensive control and monitoring capabilities; using automatic grid reconfiguration to prevent or restore outages (self-healing capabilities); enabling consumers to have greater control over their electricity consumption and to actively participate in the electricity market.

Smart grids can create a revolution in the building sector. The accumulated experience of the last few decades has shown that the hierarchical, centrally controlled grid of the 20th century is ill-suited to the needs of the 21st century. The smart grid can be considered as a modern electric power grid infrastructure for enhanced efficiency and reliability through automated control, high-power converters, modern communications infrastructure, sensing and metering technologies, and modern energy management techniques based on the optimization of demand, energy and network availability. The role of buildings in this framework is crucial. This chapter addresses critical issues related to smart grid technologies and the integration of buildings in this new power grid framework (Güngör *et al.* 2011). The main objective of this chapter is to provide a contemporary view of the current state of the art in the potential of buildings and communities to be integrated in smart grids, as well as to discuss the still-open research issues in this field. Since the vast majority of smart grids' potential customers are buildings (residential, commercial, retail and industrial) and communities, the chapter addresses the challenges posed by smart grids on the building and community level.

1.1. Smart and zero-energy buildings

The energy consumption for buildings accounts for 40% of the energy used worldwide. It has become a widely-accepted fact that measures and changes in the building modus operandi can yield substantial energy savings, minimizing the buildings' carbon footprint (Santamouris and Kolokotsa 2013; Deng *et al.* 2014). Moreover, buildings in the near future should be able to produce the amount of energy they consume, that is, become zero or nearly zero-energy buildings (ZEBs) (Kolokotsa *et al.* 2011; Pyloudi *et al.* 2015). This is a mandatory requirement based on the fact that by December 31, 2020, all new buildings were nearly zero-energy consumption buildings. New buildings occupied and owned by public authorities needed to comply with the same criteria by December 31, 2018 (Kapsalaki and Leal 2011; Kolokotsa *et al.* 2011).

ZEBs are buildings that work in synergy with the grid, avoiding putting additional stress on the power infrastructure (Li *et al.* 2013). Achieving a ZEB includes, apart from minimizing the required energy through efficient measures and covering the minimized energy needs by adopting renewable sources, a series of optimized and well-balanced operations between consumption and production, coupled with successful grid integration (Carlisle *et al.* 2009).

Information and computer enabled technologies (ICT) and smart grids implementation are the keys to achieve the aforementioned zero energy goals (Privat 2013). ICT for energy management in buildings has evolved considerably in the last decades, leading to a better understanding and usage of the term "smart buildings" (Nikolaou *et al.* 2012). Advances in the design, operation optimization and control of energy-influencing building elements (e.g. HVAC, solar, fuel cells (FC), CHP, shading, natural ventilation, etc.) unleashed the potential for the realization of significant energy savings and efficiencies in the operation of both new and existing dwellings worldwide. Smart buildings ready to be interconnected with smart grids should comply with the following requirements:

a) incorporation of smart metering;

b) demand response capabilities;

c) distributed architecture;

d) interoperability.

1.1.1. *Smart metering*

Smart metering is a prerequisite and starting point for the effective implementation of smart grids and zero-energy buildings. In Finland, the usage of smart metering encouraged consumers to increase energy efficiency by 7%. In order for electricity providers to deliver intelligent services for customers, bidirectional metering interfaces should be used to obtain customers' energy demand information (Bae 2014). Moreover, through the advances of smart metering, sensors-based approaches can be exploited to provide power load forecasting (Jain *et al.* 2014). Data collected from smart meters, building management systems and weather stations can be used by advanced artificial intelligent techniques and machine learning algorithms to infer the complex relationships between energy consumption and various variables such as temperature, solar radiation, time of day and occupancy (Mellit and Pavan 2010; Gobakis *et al.* 2011; Zhao and Magoulès 2012; Jain *et al.* 2014; Jetcheva *et al.* 2014; Papantoniou *et al.* 2016). Due to the fast development and application of low cost options for energy metering in recent years, power load prediction is becoming increasingly relevant and cost effective (Fan 2014; Jain *et al.* 2014).

Smart metering with sensor-based approaches was exploited in the framework of the Green@Hospital project (www.greenhospital-project.eu/). In this project, the outdoor temperatures and hospitals' energy demand were predicted for 4, 8, 12 and 24 hours ahead (Papantoniou *et al.* 2015). This prediction is then used for optimal control of the hospitals' air handling units, leading to an almost 20% reduction of the energy used. Other researchers exploit neural networks' capabilities for 24 h-ahead building-level electricity load forecasting, using data collected from various operational commercial and industrial building sites (Jetcheva 2014). Data mining-based approaches which collate models for predicting next-day energy consumption and peak power demand, with the aim of improving the prediction accuracy, are also developed. This approach was adopted to analyze the large energy consumption data of the tallest building in Hong Kong (Fan 2014) with very satisfactory results. These ensemble models can be valuable tools for developing strategies of fault detection and diagnosis, operation optimization and interactions between buildings and smart grids. Moreover, data processing and interpretations extracted by the smart meter can provide useful information for the buildings' energy behavior. Advanced techniques such as cluster analysis are used by various researchers (Nikolaou *et al.* 2012; Panapakidis 2014), leading to the determination of optimum clustering procedures as well as building benchmarking.

1.1.2. *Demand response (DR)*

DR (Bartusch and Alvehag 2014; Li and Hong 2014) offers the ability to apply changes in the electricity usage by the consumers from their normal consumption patterns in response to changes in electricity pricing over time (Bradley *et al.* 2013). This leads to lower energy demand during peak hours or during periods that the electricity grid's reliability is put at risk. Therefore, demand response is a reduction in demand designed to reduce peak load or avoid system emergencies. Hence, demand response can be a more cost-effective alternative than adding generation capabilities to meet the peak and/or occasional demand spikes. The underlying objective of DR is to actively engage customers in modifying their consumption in response to pricing signals. DR is expected to increase energy market efficiency and security of supply, which will ultimately benefit customers by way of providing options for managing their electricity costs, which leads to a reduced environmental impact.

The already available DR programs are generally categorized into incentive and price-based programs. Incentive-based programs provide economic incentives for customers to reduce demand at times of capacity shortage or exceptionally high electricity prices, whereas price-based demand response programs involve dynamic tariff rates that promote general changes in patterns of electricity use. Time-of-use tariffs, which are one of the major price-based DR programs in use involve different unit prices within different blocks of time, and reflect the average cost of utilities during these periods (Bartusch and Alvehag 2014).

There are some efforts at the country level that show the benefits of DR in electricity supply. The policy discussions in the UK on the economic case for DR are analyzed by Bradley *et al.* (2013). A cost/benefit analysis is performed in a quantitative manner showing that the benefits on a country level are clearly very significant, that is, there was a 2.8% reduction in overall electricity use and a 1.3% shift in peak demand. Moreover, the economic viability of the DR mainly depends on ensuring participation by the end users, that is, the building sector. An increase in participation can be ensured by lowering the participant costs and sharing of benefits. Finally, it is revealed that the actual costs of the infrastructure are also affected by customer engagement and trust. An empirical study for Sweden is performed by Bartusch and Alvehag (2014) in order to estimate the end users' response to a demand-based time-of-use electricity distribution tariff among Swedish single-family residential houses. The study showed that in the long term, the residential households still respond to the price signals of the tariff by cutting demand in peak hours and shifting electricity consumption from peak to off-peak hours.

Energy efficient smart buildings are possible by integrating a smart meter, smart sockets, domestic renewable energy generation and energy storage systems for integrated energy management, and this integrated system supports demand side load management, distributed generation and distributed storage provisions of future smart grids (Kayo *et al.* 2014; Keles *et al.* 2015). Consequently, the effective integration of buildings in smart grids requires appropriate levels of digital technology and interoperability (Oliveira-Lima *et al.* 2014). The successful implement-station of DR, as mentioned in the previous paragraphs, requires near-real-time power management (Hong *et al.* 2014) as well as advanced building automation and communication protocols. More information on the various communication protocols that can support the ZEB perspective can be found in Kolokotsa *et al.* (2011). Recent studies interconnect the ASHRAE BACnet protocol with the wireless Sigsbee protocol to provide DR in the buildings' energy management side. The addendum announced by ASHRAE specifies that the wireless communication for BACnet will be based on the Zigbee protocol (ANSI/SHRAE standard 135-2010 2013). The interconnection of BACnet with Zigbee allows the management of electrical resources and devices' load, targeting to optimize and reduce the electricity cost and reduce peak power consumption. The real-time characteristics and functionalities of the BACnet–Zigbee interconnection are examined, showing the potential and critical role of recent developments in the implementation of smart grids' infrastructure in buildings. Other researchers provide interconnection of the BACnet and EnOcean protocols to measure energy metering of various appliances in buildings and to control the load using wireless communications (Li and Hong 2014). Additionally, the DR using the specific connectivity is shown using experimental facilities that can respond to real-time changes in the electricity price.

1.1.3. *Distributed systems*

Distributed systems play a crucial part in the demand response of smart buildings, as they provide the necessary technical infrastructure (Chalkiadakis *et al.* 2011). Moreover, distributed systems support the effective exploitation of energy storage. Since smart grids may integrate renewable energy sources, energy storage is often seen as necessary for the electric utility systems with large amounts of solar or wind power generation to compensate for the inability to schedule these facilities to match power demand.

Load shifting through effective thermal energy storage or electricity storage is a major part of smart grids and already exploited by various studies for generation–consumption matching and zero energy targeting (Chalkiadakis *et al.* 2011; ANSI/SHRAE standard 135-2010 2013). Therefore, as the major power consumers on the demand side, buildings can actually perform as distributed thermal storages to help relieve the power imbalance of a grid. This requires an accurate prediction of possible power demand variations of buildings and energy information from the grid side for interaction and optimization (Xue *et al.* 2014) based on measurements acquired from smart metering. Finally, smart grids can play a significant role in the building sector due to the fact that they create a physical proximity between consumers and energy production that may help increase end users' awareness towards a more rational use of energy for buildings. Studies show that the users' awareness combined with smart grids can decrease the energy costs by 15% (Li and Hong 2014).

1.2. Smart and zero-energy communities

Moving from the building to the community level, the requests of the future communities are very demanding. They should be places of advanced social progress and environmental regeneration, as well as places of attraction and engines of economic growth based on a holistic integrated approach in which all aspects of sustainability are taken into account.

As mentioned in the Leipzig Charter (http://ec.europa.eu/regionalpolicy/archive/themes/urban/leipzig charter.pdf), sustainable communities and cities should:

– Create a suitable environment for urban populations that can attract industry, businesses, tourism and workforces.

– Modernize the infrastructure networks. This approach is about achieving a high-quality and affordable urban transport system that includes a network linking the city to the region, as well as providing and improving supply networks such as energy, waste-water treatment and water supply.

– Ensure the efficient use of natural resources, economic efficiency and the energy efficiency in new and existing buildings.

The National Renewable Energy Laboratory defined the zero-energy community as "one that has greatly reduced energy needs through efficiency gain such that the balance of energy for vehicles, thermal, and electrical energy within the community is met by renewable energy" (Carlisle *et al.* 2009) and "community scenarios could link transportation, home and the electric grid as well as enable large quantities of renewable power onto the grid".

Therefore, a smart zero-energy community (http://www. smartcommunities.org) is a region where its citizens and local authorities are exploiting the information technology to "transform life and work in significant and fundamental rather than incremental ways" to meet the aforementioned strategies and objectives. As mentioned on the Smart Communities website: "The goal of such an effort is more than the mere deployment of technology. Rather it is about preparing one's community to meet the challenges of a global, knowledge economy."

Smart grids can be the basis of smart zero-energy communities offering:

1) Reliability and security as the prominence of information technology.

2) Optimal operation, thus contributing to a generation–consumption matching (Kolokotsa *et al.* 2011) through the full exploitation of low zero carbon (LZC) emission technologies (Carlisle *et al.* 2009; Xue *et al.* 2014).

3) Adaptation to the rapidly evolving ICT technologies, as well as exploitation of the internet's capabilities, smart devices and applications (Zhen *et al.* 2012).

4) The necessary balance between the energy demand and energy production on the community, neighborhood and district level, contributing to the smart and sustainable urban environment.

Effective management of the intermittency of renewable energy sources power generation as well as of operational and capacity reserve in conjunction with the buildings' energy demand profiles. In order to understand more clearly the role of smart grids in zero-energy communities, some examples are presented in the next paragraphs. More details on smart grids worldwide popularity can be found in Bartels (2012).

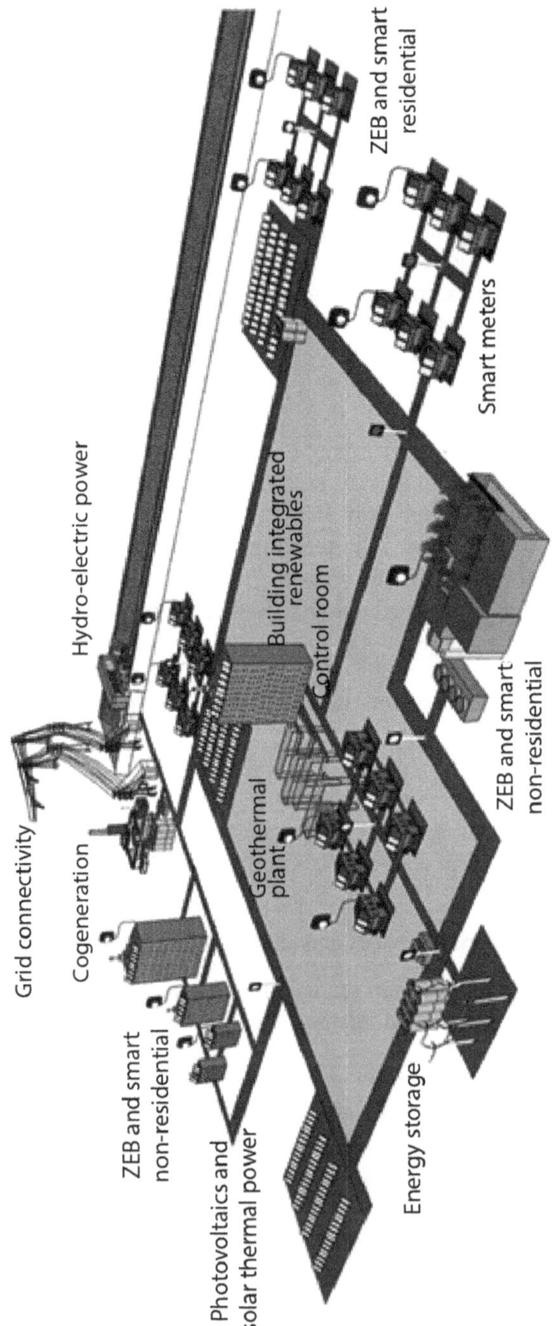

Figure 1.1. *Components of the smart grid*

The role of net energy metering and the time of use rate in the electricity demand of a zero-energy apartment community have been investigated (Gaiser and Stroeve 2014). The apartments' community is placed in West Village in Davis, California. Smart scheduling is applied in order to achieve peak load shaving by using the various household appliances during off-peak hours. The peak demand is reduced by 18%, the part-peak demand by 32%, and the off-peak demand is increased by 12%. Since the apartments are using photovoltaics for electricity, the community's occupants can benefit significantly if the surplus generation by the photovoltaics is maximized and sold to the grid during peak hours (Figure 1.1).

Moreover, a low energy community configuration for the Island of Hawaii is proposed in Zhang (2009). The overall work included energy production using photovoltaics, storage systems using batteries and compressed air systems, demand response of buildings, passive cooling techniques for energy efficiency as well as microclimate and landscape assessment.

An advanced energy management and optimization model for the operation of the Leaf Microgrid Community is proposed by Kolokotsa *et al.* (2015). The aim of the model is to minimize the energy cost of the microgrid by performing a generation–consumption matching using genetic algorithms. The optimization procedure is supported by power load forecasting and energy production prediction (by the photovoltaics and hydroelectric plant) using artificial neural networks. While the optimization horizon is 24 hours ahead, the optimization and management procedure leads to an almost 6% reduction in the energy costs for the microgrid and significant cost savings for the community without extra energy investment.

Similar works include apart from genetic algorithms (Liang and Gooi 2010; Mohamed and Koivo 2012) the use of fuzzy logic and the particle swarm optimization method (Kyriakarakos *et al.* 2012) as well as the non-linear constraint multi-objective optimization method (Mohamed and Koivo 2007).

Another potential application of smart grids at the community level is the operation of small and medium enterprises (SMEs) which have multiple operational characteristics ranging from office buildings, retail, industrial, etc. and different energy patterns and needs (Rawlings *et al.* 2014). Though the potential for carbon reduction of SMEs is unclear due to the diversity of the end uses, smart grids can be a viable tool for SME decarbonization since additional data can be available through detailed smart metering. Therefore, the electricity loads of the SMEs can be discretized and the proportions that are available for load shifting or shaving can be more easily identified.

These can be space heating, cooling and hot water in the commercial sector, and heating or cooling in industry.

Last but not least, the social aspects of smart grids at the community level should be clearly examined. The fast promotion of smart grids' capabilities and potential applications will be much easily understood by local communities if social inclusion is taken into account. The role of social experts in this aspect is of great importance (Wolsink 2012).

1.3. Conclusion and future prospects

Smart grids can be considered very promising for the energy and building environment industry due to the fact that they create a physical proximity between consumers and micro energy sources that help increase consumer awareness toward a more rational use of energy.

Moreover, smart grids can offer new opportunities for the reduction of gas emissions by creating technical conditions that increase the connection of devices and renewable energy resources at a low voltage level.

In addition, smart grids in the building sector offer a great opportunity for improving the power quality and reliability of energy sources due to the fact that it offers decentralization of supply, better supply and demand matching, reduction of transmission losses and minimization of downtimes. Thus, energy investments can be shifted from the expansion of transmission and large-scale generation systems to the energy efficiency in the building sector, that is, improving building fabric, increasing green infrastructure in the community, and improving indoor and outdoor environment interaction by using landscape solutions.

In addition, the widespread application of modular micro generation sources on a community level may contribute to the reduction of the energy price in the power market. Further, price reduction may be achieved by optimizing micro-generation operation and performing building load forecasting, which is possible due to the available data from the metering process. As a result, the smart grids can be viewed as aggregators of buildings, consumers and communities that will be empowered with better prices and valuable opportunities.

In order to reach the aforementioned goals, several challenges and drawbacks should be overcome. Technical challenges are mainly related to lack of experience, knowledge and competence necessary to successfully operate the micro generation energy sources as well as to interact with the building's occupants. These aspects require extensive research efforts with an interdisciplinary perspective that will examine smart grids' real time management, protection and control. Simultaneously, the role of the end users and building occupants should be enhanced and successfully incorporated. The Internet of Things can play a significant role in this direction as it can support raising awareness of end users and can provide user friendly applications for demand response and load management capabilities. In order to achieve this, specific telecommunication in infrastructures as well as open communication protocols should be provided to help manage, operate and control buildings within smart grids.

Another significant challenge is related to the increased initial installation cost which constitutes a great disadvantage. Specific policies and incentives should be provided by the governmental bodies to encourage investment in order to reach national carbon reduction goals.

Finally, significant effort should be put in the development of standards concerning the operational characteristics, the power quality, real time optimization and overall management.

2

Integrated Design (ID) Towards Smart Zero-energy Buildings and Smart Grids

This chapter focuses on smart and zero-energy building analysis in the design phase and aims to provide guidelines for integrated design procedures toward smart/zero-energy buildings and smart grids.

This is a mandatory requirement based on the fact that the requirements of the EPBD impose that all new buildings should meet the nearly zero-energy targets from 2021 onwards. New buildings occupied and owned by public authorities have had to comply with the same criteria from 2019 (Kolokotsa *et al.* 2011; Kapsalaki and Leal 2011).

Since buildings are major consumers in smart grids, the integrated design task helps in developing a collaborative method for designing buildings for smart grids. The IED process is an innovative approach able to support and manage the growing complexity of the building market sector. It requires multidisciplinary collaboration, including key stakeholders and design professionals, from conception to completion. Decision-making protocols and complementary design principles must be established early in the process in order to satisfy the goals of multiple stakeholders while achieving the overall integration design objectives.

The design of NZEB requires an interdisciplinary approach. Reducing the energy demand in the design phase requires the careful consideration of specifications of the different designers and engineers such as architects, building engineers or façade designers. For this reason, the introduction of a design team is compulsory for the

---

Chapter written by Theoni KARLESSI, Pietro MURATORE, Luca VENEZIA, Laura STANDARDI, Klemens LEUTGÖB and Anne Sigrid NORDBY.

For a color version of all figures in this chapter, see www.iste.co.uk/kolokotsa/smartbuildings2.zip.

design of NZEBs. In this context, the building design phase is of particular importance. IED is a valuable approach in helping to reduce the complexity of the design process, to ensure the implementation of defined targets, to identify the pros and cons of alternative variants of design concepts and to allow decision makers to decide based on transparent facts. The choices are no longer made by a single profession, but by a work team through a participatory process; choosing from a wide range of possibilities to identify the best solution, taking into account the quantitative aspects (high energy performance efficiency and high indoor comfort), economic (cost/benefit), functional, aesthetical aspects, and energy efficiency parameters to be achieved. The application of IED from the very beginning of the design phase ensures that a cost-effective solution for NZEB can be identified, because only at the early design phases changes of the general design concept can it be implemented at low cost. Therefore, the application of IED is part of the best way towards achieving the intended NZEB at low cost.

The work presented hereafter is structured in two levels:

– *Level 1: Integrated design procedure and decision making protocols*

The integrated design concept is analyzed providing the principles of the ID process and defining decision making protocols. It should be emphasized that the creative problem-solving process runs parallel to the process of monitoring the compliance with project targets, which will have been defined in the design basis. The ID steps represent concrete tasks that to a certain extent go beyond conventional design processes. This means that an adaptation of the scope of services is required. The client's and the tenant's requirements are presented, including capital cost, delivery risk, operational cost, and minimization of building unsuitability risk. The objectives of services and remuneration models are also investigated. The importance of certification and consultancy aspects is highlighted also providing quality assurance aspects and evaluation of the design quality criteria.

– *Level 2: Assessment of best practice examples through site visits*

Visits to best practice example NZEB buildings, which follow the integrated design concept from the design to the construction and operation, also integrating at the same time smart technologies, have been carried out. The buildings are documented, and their ID characteristics are presented also providing recommendations for smart grid integration.

The activities included site visits, meetings with the people responsible for the building facilities and assessment of the performance of four best practice buildings in Greece:

– A.N. Tombazis and Associates Architects Office Building;

– Apivita Commercial and Industrial S.A.;

– Stavros Niarchos Foundation Cultural Center;

Karelas Office Park.

These buildings have adopted the ID procedures from the early stages of design up until the final completion and operation.

2.1. Introduction

As the EU energy industry is clearly moving toward a new era of reliability, availability and efficiency that will contribute to the improvement of Europe's economic and environmental sustainability, smart technologies provide the opportunity for new applications with far-reaching interdisciplinary impacts: providing the capacity to safely integrate more renewable energy sources (RES), smart grids and distributed generators into the network; delivering power more efficiently and reliably through demand response and comprehensive control, and monitoring capabilities in order to achieve zero-energy targets. The integration of smart technologies requires a holistic approach that takes into account all aspects of sustainability.

The methodological approach is based on a cycle expansion of three phases (Figure 2.1):

1) the user/consumer aspects, focusing on smart and zero-energy buildings analysis;

2) the penetration of smart grids at community and city level, offering the technological platform for moving quickly toward sustainable communities by exploiting the ICT and energy systems development to the maximum degree;

3) development of optimized operation of smart applications and smart grids.

Phase 1 starts from the user/consumer aspects by focusing on smart and zero-energy buildings analysis (Carlisle *et al.* 2009). This is the core of the cycle approach.

The design of NZEB requires a holistic approach. The reduction of energy demand in the design phase is related to the involvement and setting of specifications by the whole team of experts. For the demand side concept of a building, the best possible heating or ventilation system should be applied. Thermal mass activation for example requires interaction between the structural designer and HVAC engineers. Alternative energy systems have to fit with the concept design and the building's energy systems. For this reason, the introduction of a design team is compulsory for the design of

NZEBs. In this context, the building design phase is of particular importance. IED is a valuable approach in helping to reduce the complexity of the design process, to ensure the implementation of defined targets, to identify the effectiveness of alternative choices, and methods of design concepts and to allow decision makers to implement the appropriate solutions based on transparent and solid methodologies. Only if IED is applied from the very beginning of the design phase, is it highly probable that a cost-effective solution toward NZEB can be identified, because only at the early design phases can any deviation from the general design concept be implemented at low cost. Therefore, the application of IED constitutes the best way toward the intended NZEB at low cost. Experience from several demonstration projects and best practice examples shows that IED leads to highly energy efficient solutions at least cost over the lifecycle of the building, because the integration of all the required expertise even in the early design phase brings forward the most appropriate and thus most cost-efficient solutions (Nordby 2013).

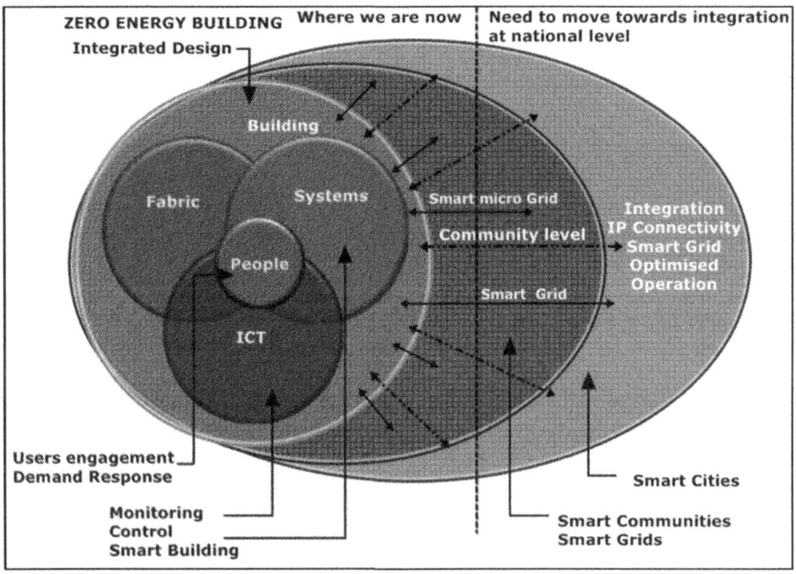

Figure 2.1. *The methodological approach of smart technologies*

2.2. Methodology

In order to provide a holistic approach, the following methodology has been implemented (Figure 2.2):

– *Integrated design concept* is analyzed providing the principles of the ID process and defining decision making protocols. Client's and tenant's requirements are presented, including capital cost, delivery risk, operational cost and building unsuitability risk reduction. The objectives of services and remuneration models are also investigated. The importance of certification and consultancy aspects is highlighted, also providing quality assurance aspects and evaluation of the design quality criteria.

– Site visits to *best practice example NZEB buildings*, which follow the integrated design concept from the design to the construction and operation, integrating at the same time smart technologies, have been carried out. The buildings are documented, and their ID characteristics are presented also providing recommendations for smart grid integration.

– *Case buildings were fully assessed* and the methodology that was developed included the audit and analysis of the building's existing situation.

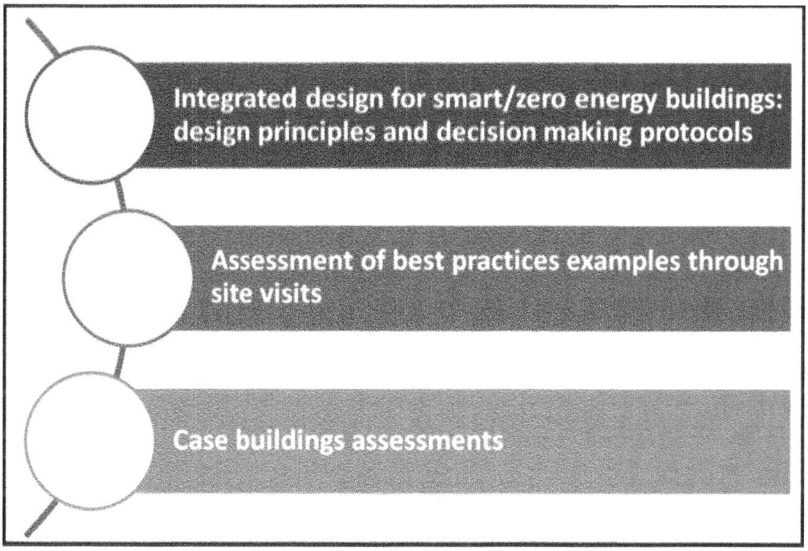

Figure 2.2. *The methodology of implemented work*

2.3. Integrated design in smart and zero-energy buildings

Directive 2010/31/EU (EPBD recast) Article 9 requires that "Member States shall ensure that by 31 December 2020 all new buildings are nearly zero-energy buildings; and after 31 December 2018, new buildings occupied and owned by public authorities are nearly zero-energy buildings". Member States shall furthermore "draw up national plans for increasing the number of nearly zero-energy buildings" and "following the leading example of the public sector, develop policies and take measures such as the setting of targets in order to stimulate the transformation of buildings that are refurbished into nearly zero-energy buildings".

A nearly zero-energy building is defined in Article 2 of the EPBD recast as "a building that has a very high energy performance. The nearly zero or very low amount of energy required should be covered to a very significant extent by energy from renewable sources, including energy from renewable sources produced on-site or nearby". The specific EPBD Concerted Action activities around "Towards 2020 – Nearly zero-energy buildings" will support the Member States by the exchange of experiences with already existing high-performance buildings (ranging from low energy buildings to passive houses, zero-energy and zero-emission buildings, and even to energy surplus houses). The discussion topics include the most common building and service system solutions, calculation methods, promotional means, available subsidies and other incentives, supporting documents (e.g. guidelines), etc. as well as study tours to interact with experts at national administrations and visits to relevant sites.

There are different national applications of the definition of nearly zero-energy buildings: in some countries NZEBs concept is well established and they act as best practises examples providing valuable experience to other that are at a level of development of this concept. Through such information exchange, Member States participants are mutually supported in the development of the national plans for increasing the number of nearly zero-energy buildings. To achieve a suitable definition, related facts, findings and experiences need to be seen in a broader societal context and need to be transferred into a well-defined standard, taking into account financial, legal, technical, and environmental aspects. Analyzing the implications identified above, it becomes obvious that most of them interact or require the consideration of a holistic approach (Karlessi *et al.* 2014).

Consequently, the principles for an NZEB definition should be based on the same broad perspective, addressing the present and future challenges and benefits. Hence, a proper and feasible NZEB definition should integrate the following characteristics (Thomsen *et al.* 2011):

– to be clear in its aims and terms, to avoid misunderstandings and implementation failures;

– to be technically and financially feasible;

– to be sufficiently flexible and adaptable to local climate conditions, building traditions etc. without compromising the overall aim;

– to build on the existing low-energy standards and practices;

– to allow and encourage open competition between different technologies;

– to be ambitious in terms of energy and environmental conditions considering innovativeness and technology development;

– to be elaborated based on the agreement of the main stakeholders involved (politicians, designers, industry, owners, investors, users etc.);

– to be inspiring and to enforce faster adoption, acting also as an example for other projects.

Within this context, one main objective is the design of common projects concerning smart buildings in an interdisciplinary nature in order to accelerate the process toward zero-energy buildings through integrated design.

2.4. ID process principles and guidelines

Integrated design is an approach that considers the design process and the physical solutions as a whole system, which provides optimized buildings throughout their lifecycle. First, for the purpose of reaching a high sustainability performance, the project should be developed and discussed by an integrated, multidisciplinary team.

ID emphasizes a decision process rooted in well based choices with regard to the project goals, and on the systematic evaluation of design proposals at every stage of the development. This approach to building design parallels the principles of environmental management referred to in the international ISO 14001 standards. Here, the identification, definition and prioritization of goals, as well as the development of an evaluation plan with milestones, are central and critical issues.

The concept of an integrated approach emphasizes that the very early phases need more attention because well informed decisions here will pay off in the following stages of the design process and eventually in the lifecycle of the building (Figure 2.3) (Nordby 2013). The IED approach is an effective way to realize NZEBs because it involves different specialties interacting on energy performance issues during the early stages of the design process. Figure 2.3 shows the difference between a traditional approach (gray area) and an IED process (orange area). With the integrated design process, the design phase requires more effort than the construction and documentation phase. At the same time, the cost's curve trend changes with the decision making time phase, in the IED approach it is high during the design phases (blue line) while in a traditional approach (green line) it is high during the construction and operation phase due to project changes (Leutgöb 2012; Nordby 2013).

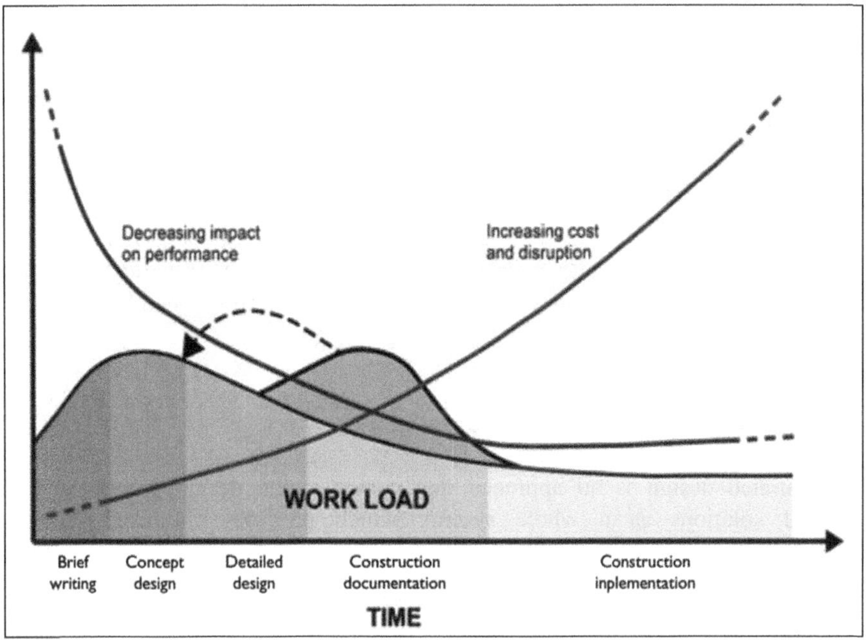

Figure 2.3. *Early design phases impact on performance, costs, and disruption (www.integrateddesign.eu)*

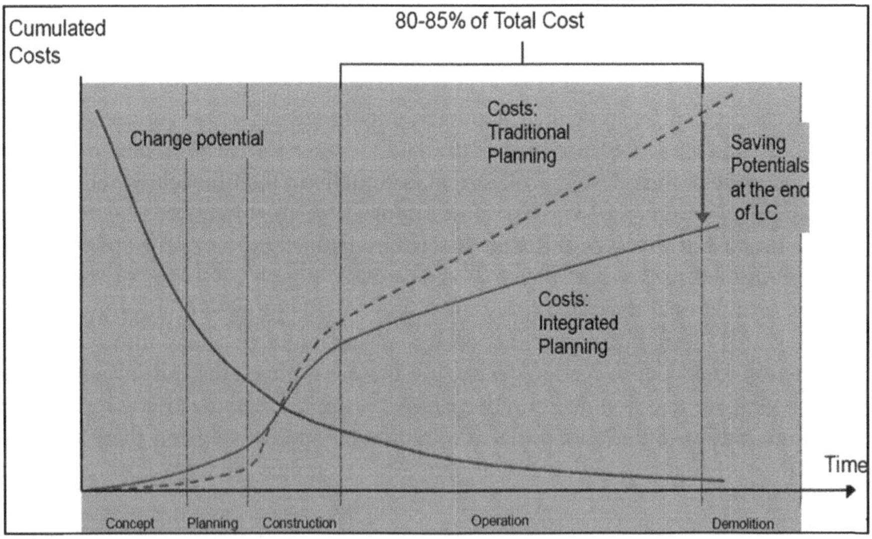

Figure 2.4. *Traditional versus Integrated Planning*

Well informed planning at the beginning of the project can allow buildings to reach high energy performance and reduced operating costs at minor extra capital cost. Considering the whole lifecycle of a building, the running costs are higher than construction and refurbishment costs; thus, it becomes obvious that a short-sighted approach to squeeze the first design phase regarding resources should be avoided. Experience from building projects applying ID shows that the investment costs may be about 5% higher, but the annual running costs will be reduced by as much as 40–90%. The process of ID emphasizes that the performance of buildings should be assessed in a lifecycle perspective, both regarding costs (LCC) and environmental performance (LCA). Figure 2.4 indicates the cost effectiveness of Integrated Planning versus Traditional Planning through the lifecycle of the building.

Six steps can be identified for a successful Integrated Design implementation (Nordby 2013; Leutgöb 2012):

1) *Project development*: discuss the project ambitions and challenge initial client presumptions, initiating ID process, and preferably make partnering contracts.

2) *Design basis*: select a multi-disciplinary design team, including an ID facilitator, motivated for close operation, analysis of the boundary conditions. Also define objectives and specify the project ambitions, preferably as functional goals.

3) *Iterative problem solving*: facilitate close cooperation between the architect, engineers and relevant experts through workshops, seminars, etc. Use of both creative and analytical techniques in the design process. Discuss and evaluate the multiple concepts and finalize optimized design.

4) *On track monitoring*: use goals/targets as means of measuring success of design proposals, make a Quality Control Plan, evaluate the design and document the achievements at critical points/milestones.

5) *Delivery*: ensure that the goals are properly defined and communicated in the tender documents and building contracts, motivate and educate construction workers and apply appropriate quality tests, facilitate soft landing. Make a user manual for operation and maintenance of the building.

6) *In use*: facilitate commissioning and check that the technical systems, etc. are working as assumed, monitor the building performance over time regarding, for example, energy consumption, user satisfaction, etc.

This procedure is depicted in Figure 2.5. The figure emphasizes that the creative problem-solving process (2) runs parallel in time with the process of monitoring (3) the compliance with project targets, which have been defined in the design basis (1).

The ID steps represent concrete tasks in a circular iterative way rather than a linear approach. That to a certain extent goes beyond conventional design processes. This means that an adaptation of the scope of services is required.

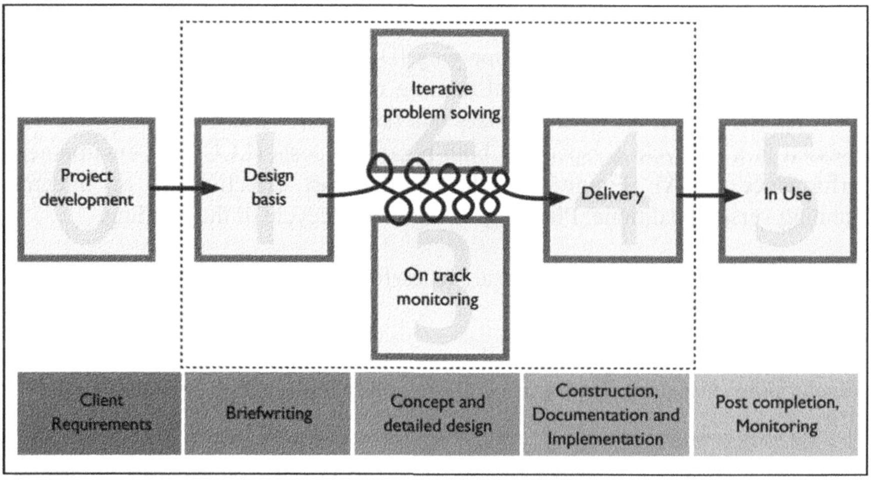

Figure 2.5. *Overview of the ID process*

2.4.1. *Benefits*

Integrated Design processes result in higher energy and environmental performance: all related building aspects are addressed through open multidisciplinary discussions and design decisions in early project phases, where knowledge and experiences are exchanged to provide the most effective solutions.

It also contributes to the reduction of CO_2 emissions as optimized design is prioritized combined with advanced technical systems and control mechanisms.

The indoor climate is significantly improved: the building and technical systems work together in a logical symbiosis in order to achieve sufficient indoor air quality, temperature control and daylight access/solar protection.

Running costs and maintenance of the building are reduced: efficient technical systems are more cost effective, both in terms of initial investment costs for manufacturing and installation and in terms of running costs and maintenance.

An important benefit is the reduction of risks and construction defects as improved planning leads to fewer building faults. This results in significant financial savings. Early involvement of users and inclusion of their requirements in the design process may improve the overall performance of the building in the operation phase, as well as increase user satisfaction. A high-performance building increases the attractiveness and can yield higher rental costs which can be compensated for by a lower energy bill, thus the sales value of the building will increase. A green image also enhances the building profile and benefits the building owner or tenant organization.

2.4.2. *Barriers*

Despite the important benefits, there are certain barriers that need to be addressed, such as:

1) Conventional thinking: The building sector is considered to be rather slow and reluctant in accepting and adopting new ways of working. ID calls for decision processes and design methods that challenge familiar habits and require high communication skills. Professionals on both sides of the table must practice in collaboration, and maybe adjust their working habits.

2) ID does not seem a cost-effective approach: Developers traditionally pay more attention to construction costs than lifecycle costs (LCC). However, when energy consumption and maintenance is included in the calculations, it usually supports investments in planning for high performance and robust solutions.

3) Time constraints in the initial design phase often press engineers and developers to underestimate the value of thoroughly and integrated planning and propose high speed and inappropriate solutions when analyzing the boundary conditions of a building. It can be challenging to convince the developer that the initial phase is crucial, and that giving time for design iterations often pays off in better concepts and later phases.

4) As the ID process requires close and constant collaboration between stakeholders who may have diverging goals, conflicts could be accentuated. It is therefore crucial that the team members do not insist on strict and definite demands within their fields of expertise, but rather endeavor to work with a holistic approach.

The inclusion of a "design process facilitator" should be considered, in order to organize the process, especially in cases where the architect and client lack knowledge of collaborative working or where the project has particularly challenging performance goals. It is highly recommended that he/she should be assigned and contracted separately to the client in order to guarantee effective coordination and management of the ID process and to avoid the resolve of issues of private interest. Supervision and on-track monitoring throughout the design process should be performed by a facilitator or other person with the authority and experience to challenge both the design team and the client. This facilitator should also be responsible for reporting to the client any variations against the project goals originally defined. A process facilitator such as a BREEAM or LEED Accredited Professional (AP) may be the right option in cases where environmental goals are pursued through an environmental assessment scheme.

2.5. Scope of services

One basic element of ID, which is not common in conventional design processes, is the requirement of a comprehensive evaluation of the multiple concepts and technical solutions prior to taking definite decisions. This requirement implies that design work needs to be applied in several projects because a comprehensive evaluation can only be implemented if the concepts and technical solutions in question are available in a sufficiently advanced design.

A principal characteristic of ID is an upstream shift of workload from detailed design and construction documentation towards the phase of concept design. This characteristic also has an impact on the scope of services. The importance relies not only on the certain tasks that will be implemented, but also on the timeframe of the implementation. Dynamic thermal simulations, estimation of indoor conditions, and daylight calculations have become a standard tool for the design of complex buildings over recent years. Very often these procedures are applied too late in the process, acting not as evaluation of the process, but with the main aim of proving that the selected design variants work. From an ID perspective, these instruments need to be used as decision-supporting tools for filtering out the most suitable variants. This purpose can only be achieved if they are applied very early in the design process and if they include the comparative analysis of a broader range of design variants; the same accounts for lifecycle cost assessment (LCCA). Considered from an ID perspective, an estimation of the future costs, which are usually calculated rather late during the detailed design phase, is not adequate. Instead, ID uses LCCA to compare different design variants from an economic point of view and this kind of comparison is only useful if its results can be fully taken into account without causing larger disruptions in the design process as a whole. Thus, it is concluded that the LCCA needs to be applied at the very beginning of the design process. Figure 2.6 depicts the series of the most important typical ID-related tasks in the design phases and when they should be implemented, or at least initiated.

The accurate timing of tasks is crucial in order to fully reclaim the benefits of ID (Leutgöb 2013).

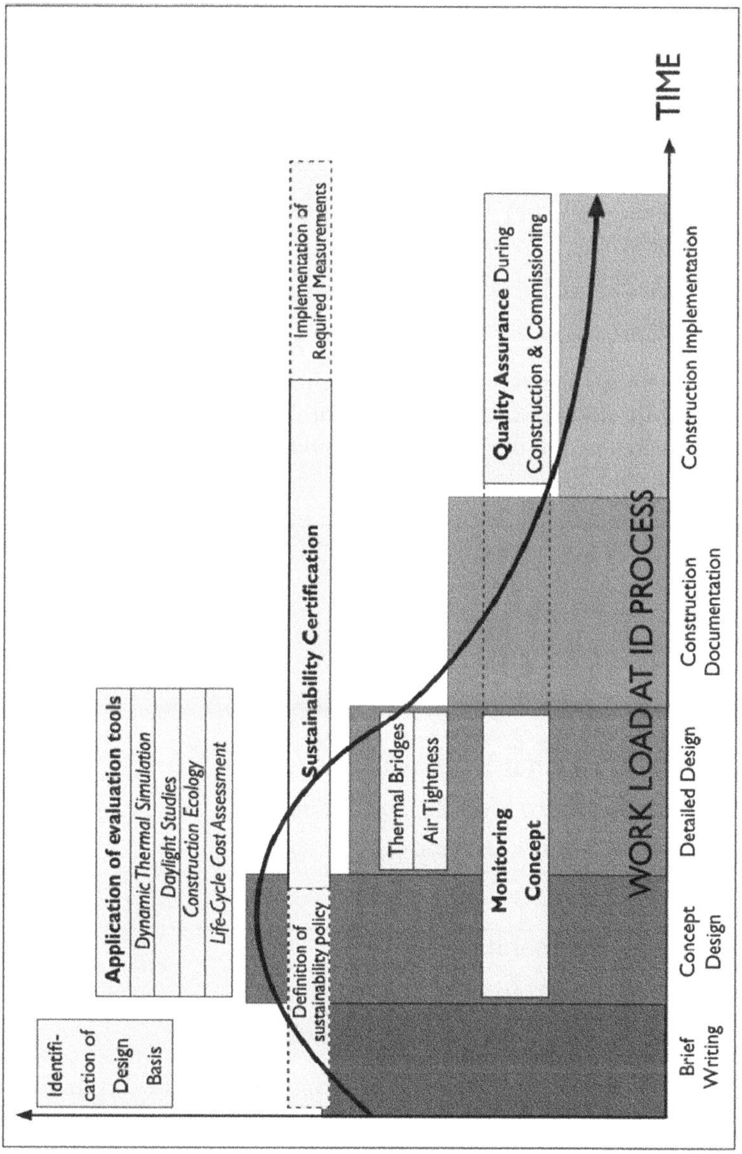

Figure 2.6. *Timeframe of ID related tasks (www.integrateddesign.eu)*

2.6. Remuneration models

The impact of ID on the remuneration of design services can be divided into four aspects:

– qualitative aspects resulting from the fact that quality targets (e.g. energy performance, sustainability targets, etc.) are the usual targets of ID processes;

– quantitative aspects arising from extra tasks which are not part of a conventional design process;

– time-related aspects reflecting the fact that ID is shifting design efforts toward the early design phases;

– structural aspects considering that ID introduces an ID facilitator as an additional stakeholder.

The first two elements are reflected in a generic approach for an ID-related remuneration model for design works, which we call the three-level-model because it states that remuneration for design works committed to ID consists of three different elements (Figure 2.7).

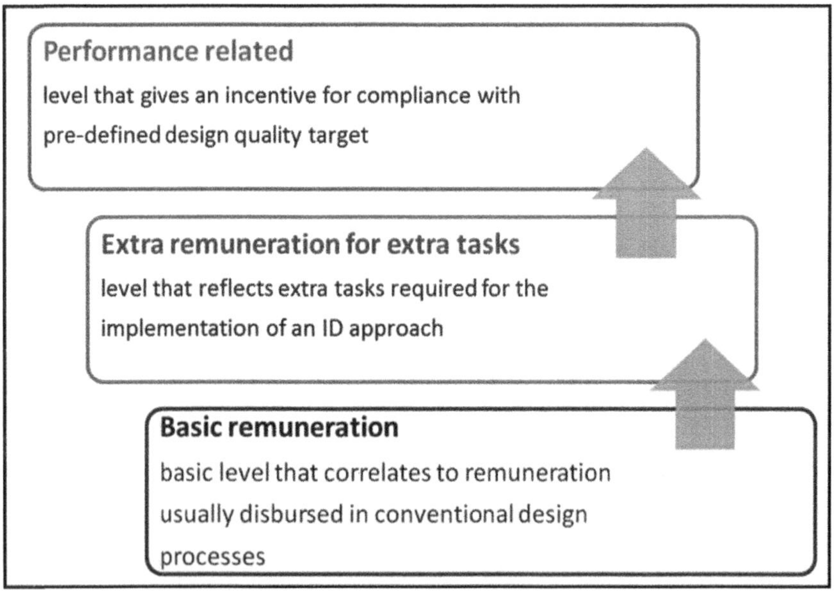

Figure 2.7. *The three-level-model for ID-related remuneration of design works (www.integrateddesign.eu)*

Level 1 – Basic remuneration

The basic remuneration is in accordance with the standard remuneration for conventional design processes. From country to country – but also from client to client – this "standard remuneration approach" shows considerable differences. In principle three different approaches are possible and may be blended in different ways:

– remuneration based on a certain percentage of construction cost;

– time-dependent remuneration (e.g. for force account works, which are added later to a running contract);

– an all-inclusive price which is fixed before the design works start.

Level 2 – Extra remuneration for extra tasks

This level of remuneration reflects the additional tasks which are not necessarily implemented in a conventional design process, but which are required for ID. If the basic remuneration is not sufficient to cover these tasks they need to be disbursed separately. This might be accomplished either by granting an extra percentage of the construction cost or by foreseeing additional time-dependent actions or fixed remunerations for the accomplishment of specific tasks.

Level 3 – Performance-related remuneration

The third level of remuneration reflects the ambition of ID to contribute to the improvement of design quality. With respect to remuneration this ambitious goal could be supported by introducing an incentive for the designer to address the quality target defined at the beginning of the design process. The way of measuring the design quality is the main challenge of this process. Despite the difficulty of these aspects, a limited share of performance-oriented remuneration seems to be feasible, since several quality criteria and performance indicators can be checked by well-established verification methods and partly also by neutral verification bodies (e.g. sustainability certificate at the design stage). The performance-related part of remuneration tries to enforce the ID procedures and strengthen the attention of design processes on quality aspects – such as energy performance, user comfort and a broad range of other sustainability criteria – contrary to the linear ordinary process and the omnipresent focus on compliance with construction costs.

The three-level-model, as presented above, is seen as a generic approach which needs to be adapted to national requirements such as the fee structures for architects and engineers, which are settled by the respective professional bodies in many countries. Existing legislation also needs to be taken into account. In addition, the approach has to be modified for different organizational models of design and construction works – such as separate contracts for different design tasks, general planner approach, and general contractor approach. Altogether, it has to be underlined that ID is not necessarily more expensive than conventional design, even if the three-level-model might imply this. The focus on early design phases will have a clear cost-cutting impact later on. It can be assumed that this impact exceeds the extra costs of the early design phases.

Typically, there are several additional services arising from the ID process related to the identification of the design basis conditions, the application of evaluation tools, the sustainability verification, consultancy and quality assurance issues.

2.7. Application of evaluation tools

The appropriate application of evaluation tools requires an analysis of the design basis conditions identifying the aspects that should be considered:

– identification of the client's sustainability policy and requirements: this can be done by considering sustainable design schemes (Passivhaus certification scheme, BREEAM, LEED, DGNB, ITACA etc.);

– evaluation of the local climate and analysis of the prevailing meteorological conditions during summer/winter and day/night). Investigation of the preconditions influencing building orientation, surrounding buildings, obstacles, sun angle, shading, micro-climate, daylight availability, outdoor air quality, noise sources, acoustic intensity etc.;

– determination of relevant boundary conditions for building concept regarding thermal, visual and acoustic comfort, indoor air quality, operational and embodied energy etc.;

– integration of (preferably on-site) renewable energy sources: location, availability of natural resources, climate, studies on the feasibility of integrated geothermal systems.

The following evaluation tools may be part of iterative problem solving and/or part of on track monitoring. They are aimed at supporting the design team in filtering out the optimal design variant by checking to which degree different variants fulfil the defined project goals and targets.

1) Dynamic thermal simulation

The dynamic thermal simulation of a building or reference zones is frequently used in order to assess various issues such as:

– indoor temperature and humidity ranges, as well as hourly heating, cooling, and power loads for building operation;

– different design alternatives, solutions and scenarios, building forms and functions;

– energy performance calculations: demand and supply, energy required for heating, cooling, humidification, dehumidification, ventilation, pumping, artificial lighting, electric equipment;

– indoor conditions assessment: thermal comfort estimation based on calculation tools;

– determination of shaded façades: sun study, optimization of shading device;

– project-related climate studies: insolation, total available solar energy and average temperatures, Mollier chart (temperature-humidity), wind rose.

2) Calculation of benchmarks

The calculation of benchmarks refers to ecology aspects of the construction (different kinds of material-related indices, many of which are now defined only on a national or regional basis): thermal and optical characteristics of materials such as solar reflectance and infrared emittance, index of global warming potential, index of photochemical ozone creation potential, index of acidification potential and others.

3) Lifecycle cost assessment

This is the economic evaluation of different variants to identify cost optimal solutions over the lifecycle of the building (as complementary information to construction cost). LCCA is especially useful when project alternatives, that fulfil the same performance requirements but differ with respect to initial costs and operating costs, have to be compared in order to select the one that maximizes net savings.

2.8. Sustainability certification

If the sustainability policy for the project has been set on the basis of sustainability certification schemes (e.g. BREEAM, LEED, DGNB, ITACA), its achievement has to be checked at certain checkpoints by assessing the fulfilment of single criteria. Depending on the sustainability certification scheme selected, this refers to criteria in the following fields:

– *Management*: project brief and design, LCC, commissioning and handover.

– *Location and transportation*: accessibility, land protection and facilities.

– *Sustainable sites*: construction activity pollution protection, environmental site assessment, heat island reduction, tenant design and construction guidelines.

– *Water efficiency*: water use reduction, water use metering.

– *Energy and atmosphere*: reduction of energy use and carbon emissions, energy efficient equipment, advanced energy monitoring, demand response.

– *Indoor environmental quality*: addresses design strategies and environmental factors – such as air quality, lighting quality, acoustic design and control over surroundings.

– *Materials and resources*: lifecycle impacts, material efficiency, waste management, recycling.

– *Health and well-being*: visual comfort, indoor air quality, thermal comfort, acoustics, safety and security.

– *Innovation in Design*: finding new, innovative features for buildings, and going above and beyond sustainable building practices and strategies.

These schemes envisage an accredited person in charge of the assessment. Besides an informal check by the design team, most sustainability certification schemes offer the possibility to conduct a formal pre-check by issuing design certificates at different stages of the design process.

2.9. Consultancy and quality assurance

Consultancy regarding an optimized building design and overall performance refers, for example, to the building envelope, thermal quality, façade openings, solar radiation with respect to the local climate conditions and focuses on the detailed design phase.

– Thermal bridge calculation: minimization of condensation risks, minimization of thermal losses.

– Ventilation and air tightness: critical points.

The procedure includes the verification of the results achieved with regard to the quality criteria set for the project (sustainability policy, user comfort, etc.) through:

1) measurements: blower door test, thermography etc.;

2) user survey coupled with comfort monitoring;

3) indoor quality monitoring of installation and setting of technical systems;

4) energy monitoring: energy data analysis aiming at the detection of shortcomings in building operation.

Other actions within this framework are:

1) work sessions with the design team for general interdisciplinary issues and guidelines;

2) work sessions to develop the specific targets of energy performance and environmental concepts for design issues, such as:

a) heating, cooling, humidification, dehumidification and ventilation,

b) daylight and solar control,

c) indoor climate control,

d) energy supply strategy (e.g. co-generation, solar heating and cooling, PV, etc.),

e) energy monitoring concept: smart meters and systems,

f) use of materials,

g) water saving,

h) outdoor environment and landscape.

2.10. Measurement of design quality criteria

In the context of remuneration, the quantifiable performance criteria that can be applied are:

1) *Criteria for energy performance*: different benchmarking at each level of the ID process can be applied as a measure of the energy performance, for example net heat demand, net cooling demand, final energy demand, primary energy demand, etc. The performance level requires adjustments and clarifications, as well as the verification methodology (e.g. a standardized calculation methodology such as a building simulation).

2) *Criteria for environmental quality*: sustainability criteria are reflected by sustainability certificates. Thus, sustainability certificates may also be used to define and measure quality. The quality target "LEED Gold" can be measured by applying for a design certificate for LEED; since the sustainability certificates include a wider range of different topics and criteria, the client might wish to select specific focus areas within the framework of sustainability certificates. The achievement of the

(sub-)targets defined for these focus areas (such as ecological materials or recyclability) can be measured by applying the verification methodologies prescribed by the selected sustainability certificate.

3) *Criteria for lifecycle cost*: the usual criterion for economic performance and evaluation in the construction sector is compliance with target values on construction cost. From an ID point of view the construction cost target could or should be complemented by a lifecycle cost analysis. In principle, both values are measurable, although in practice a very clear definition of the verification methodology is necessary.

The aspect of measurability of the performance criteria for design quality – as described above – is closely linked to the timeframe of measuring the performance of the design team. Obviously, there is no best solution, but each point in time has advantages and disadvantages in regard to the measurability of design quality so an optimum method can be selected. Table 2.1 provides a short overview of the criteria.

Timeline	Measurability	Advantages	Disadvantages
After concept design and detailed design respectively	Verification only through calculated quality criteria	Design quality (compliance with design targets) depends prevailingly on the design team	Several quality criteria are not measurable at this stage
After construction documentation	Several quality criteria are already closer to reality	More quality criteria are measurable than in the earlier design stages	Increasing influence of suppliers on the quality by submission of tender materials
After construction implementation	Some quality criteria can be measured directly already	Clearer and more distinct picture of the final quality of the project compared to the predefined quality targets	Execution of construction work introduces an additional impact on quality
After a certain period of operation	Operation-related quality criteria can be directly measured	Selected quality criteria can undergo a reality check (e.g. energy consumption, comfort parameters, etc.)	Besides design quality, quality of execution of construction work and operational quality have an important impact on the overall result

Table 2.1. *Overview of the measurability criteria of ID*

The assessment of quality of design work is a complex challenge that offers no simple or obvious solution. Although in theory it is possible to measure certain design-related quality aspects at different points in time, using performance indicators, etc., there is hardly any experience of integrating performance aspects into the designers' remuneration.

On the other hand, ID puts forward quality targets and therefore a reflection of this aspect in the remuneration of the designer would be a logical step. Therefore performance-related remuneration elements are proposed to become part of the generic approach for an ID-related remuneration model. Taking the complexity of the topic into account, it is clear that performance-related elements are not suitable to become a major part of the designers' remuneration, but they may serve as extra incentives for performing work better than standard. Bearing in mind this limitation, we think that performance-related parts of remuneration for designers are feasible.

2.11. Defining a client's objectives

There are no obligatory requirements referring to the ID process so each client can choose to approach the aims of Integrated (Energy) Design in a way that best suits their priorities. However, a typical framework could be adopted, highlighting both the iterative procedure and the additional methods available.

The key to ID as stated before is the appointment of a core design team from the start of the project who are instructed to work collaboratively and openly to satisfy the client's objectives and ensure an environmental and energy-efficient building operation. Clients should therefore ensure that the core team are appointed at the initial stage of the project.

All information related to the site's environmental factors should be gathered by, or for, the design team, and presented in an easily understood format to all parties. Clients should therefore ensure that appropriate surveys, data collection and reports are both budgeted for in the initial development appraisals and commissioned as early as possible. Inevitably, the client expectations for the site (especially if the land/building has already been acquired or there is a potential tenant) are the basis of the procedure. This step determines the brief and gives the design team the opportunity to review elements that may be restricting the optimum design or cost outcome. To this end, an open forum meeting or workshop involving all the design team, with the initial client brief circulated in advance, and with an invitation to feedback at the meeting, is essential. Clients should ensure that time and budget is allowed for this stage, and where a future tenant is known, they could be involved in the process. The second part of this step is to translate the agreed brief into clear targets for the design team to deliver through the design stage, construction and with the completed project. Where possible, these clear targets should be contractually embedded. An obvious difference between ID and traditional design development is that, for concept designs, the design team should be required to present several alternative proposals in an iterative way. This helps the reduction of the risk of a single solution becoming overestimated whilst ensuring different approaches are explored. It is important for the effectiveness of the process that clients ensure the engagement of any known tenants, and that the design team's appointments reflects the requirements, and moderate time and costs, for undertaking this approach. The series of meetings and workshops, together with a requirement for multiple early concept designs that are assessed both creatively and analytically, are the tangible differences of an ID approach versus conventional approaches. The final workshop session for the design team and client should select the optimum design solutions addressing the client's requirements. It should be ensured that the core team presents all the information needed for this meeting, and that any known tenants or building end-users are present to input into the final selection of the scheme to be developed. Upon the selection of the final design, an updated version of the client's brief and targets must be generated to accurately represent what is intended to be delivered: where tenants are present such updates may also be required for any Agreement to Lease. This updated brief and associated goals can be used to form contractual obligations, incentive payments or delivery metrics for both the design team and contractor defining the financial aspects of the project. The client objectives should directly feed into the development outline and detailed specifications and the specific targets reflected in these documents. As a successful project implementation scheme, ID is reliant on communicating the critical issues, notably the client brief and targets, through the whole construction team delivering the building work. This should include embedding the ID goals into all relevant and contractual documentation, but also plans to communicate them to all involved stakeholders. Another aspect that could be taken into consideration is site training in relevant skills. ID is also efficient on communicating the presumptions and optimum usage methods to the end users of the project as well. In this regard, Building User Manuals, "Soft Landing" contracts, and early engagement with the building management personal are important.

2.11.1. *Capital cost reduction*

Adopting an ID approach to design can be an effective measure to reduce the overall capital cost of the project. However, this overall cost reduction is realized in a later phase of the project, and inevitably the initial stages of design work will result in slightly higher costs due to the extra work effort at the early design phases whilst the core design team develop the optimized scheme design. The slight increase in core design team costs are offset through the improved design delivered through the IED process. Typically, an IED approach will add 0.2% to the project costs before planning, but deliver 3% cost saving on the overall capital spend. Cost reductions are realized through a more coordinated and rationalized design that is better understood by the core team, with potential for notably better interaction between the architectural envelope and internal services, often yielding some of the largest savings. The process of ID can also operate with the adoption of building information modeling (BIM). BIM is a coordinated set of data about the design (often, but not necessarily, in 3D) which is developed by the whole design team, and as such fits with the collaborative approach of ID.

2.11.2. *Delivery risk reduction*

The collaborative design approach embodied within ID reduces the risk inherent in building design and construction processes through better coordination at early stages in the design development, and a better understanding of the underlying design principles by all key parties through improved communication and collaborative working. This risk reduction is complex to quantify, even through the trial ID projects, since it would require identical "non-ID" projects to be coupled for comparative analysis. However, the risk reduction leads to a reduction of problems and deviations which could otherwise result in an increase in project construction budget as well as those which could entail delays to the project implementation.

2.12. Defining the tenant's objectives

Construction projects are frequently formed based on anticipated future space requirements identified by companies; such companies then enter into Agreements to Lease (pre-lets) with developers who, as experienced construction clients, then design and construct buildings in order to accommodate the company's requirements. At the completion of the building, the company then occupies the space created at previously determined rental rates. Typically, the building asset would also usually be sold by the developer (if not already) to an investment portfolio with other stakeholders. While this remains a proven delivery mechanism, it does however contain a fundamental dysfunction between the future tenant's requirements and the developer's (client's) motivations and initial targets; the tenant will typically be driving for low maintenance and operational costs and insist on solutions fitted to their needs, whereas the client will be proposing low capital construction costs and "standardized" solutions acceptable to a wide number of tenants. However, this proposal might neglect the specific needs of the project. The usual approach to address this issue is through the specifications and requirements embedded in the Agreement to Lease, and for larger schemes a degree of ongoing monitoring through the design development by an appointed team. Many of these specifications will include the principles of collaborative work even though such references are not adequately or clearly defined in the contract. The ID process provides a clearer mechanism for ensuring the tenant's requirements, which are also highly considered throughout the design process. This also requires the use of ID by clients as part of Agreements to Lease which provides an improved outcome for the tenant.

2.12.1. *Operational cost reduction*

The ID approach is fundamentally intended to assist design teams to ensure that schemes are designed to satisfy the defined objectives whilst using the minimum building operational energy. This is achieved through the creation of multiple design concepts and assessing their energy performance. The engagement with this process envisages that the tenant is able to directly steer the developer client and retained design team toward lower operational cost buildings. It is acknowledged that the developer client will incur marginally increased design costs for this process, which in result will finally be passed on through future rental to the tenant. This slight increase in core design team costs is typically more than offset in the construction process through the savings of the proper construction solutions, coordination and efficiency. The adoption of ID can therefore be considered, at worst, "cost neutral" for a developer client. Tenants should therefore resist any developer clients that attempt to add cost as a requirement to use an ID approach.

2.12.2. *Building unsuitability risk reduction*

The ID approach provides a collaborative design scheme starting with a detailed analysis and refinement of the developer objectives, which for the tenant is likely to be directly or indirectly derived from the intended specifications in the Agreement to Lease. This provides an initial opportunity to refine the original briefing document and allows the tenant to gain the benefit of the design teams' construction experience to review, and challenge, their requirements in order to provide the optimum solution. This interactive process provides benefits for the tenant by allowing the opportunity to improve the initial specification and plan based on the design team's experience (but not the obligation), as well as ensuring the design team have a thorough understanding of the tenant's needs in a more explicit way than can be achieved through a written and approved document. Consequently, the ID approach can provide an effective scheme tailored to the tenant's and client's needs by a combination of the requirements being better quantified and better understood by the whole design team.

2.13. Best practice sites

2.13.1. *Alexandros N. Tombazis and Associates Architects S.A. office building*[1]

The building is located in Polydrosso in Athens, Greece (Figure 2.8). It is a residential area, in the northern part of Athens, with relatively low traffic. The premises host the offices of Alexandros N. Tombazis and Associates Architects S.A. The building is one of three office buildings, which are designed as one complex around a common courtyard. Apart from the architectural practice, the complex hosts a contracting company and different engineering companies (Figure 2.9). One of the main issues was saving energy and enhancing comfort and architectural quality. The building was constructed during 1990–1995 and the design as well as the construction was a leading example in the field of sustainability and integrated design in Greece.

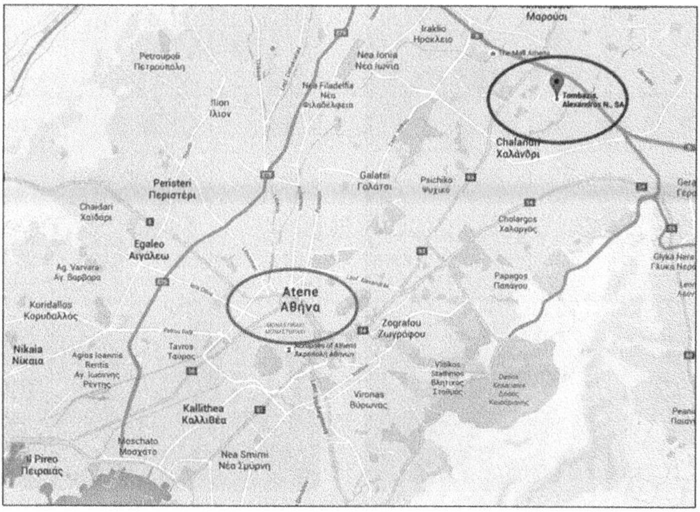

Figure 2.8. *Location of Tombazis office building (red) and city center (blue)*

1 Material for this section has been provided by kind permission from A.N. Tombazis and Associates Architects S.A., Associates: E. Kanga, Y. Romanos, O. Stefanopoulou-Plataniotі, D. Tombazi.

The building is located in a residential area in the municipality of Maroussi, Athens. The block consists of three office buildings from which the east building is the A.N. Tombazis and Associates office building.

Figure 2.9. *a) The block of buildings that includes the Tombazis building (left). b) Entrance to the Tombazis building (right)*

Figure 2.10. *Artificial pool, sculpture, and entrance to the Tombazis building*

The total building area is 1,100 m^2. The design is developed in three levels, as well as a basement with a parking area. The two long sides of the office building face east and west. For reasons of bioclimatic design, the building is narrow and long, approximately 8 x 35 m. Thus all workspaces are adequately supplied with daylight through side windows. As far as was possible, the mature olive trees that existed on the site were saved or transplanted. An important feature of the landscaping is the pond, a sculpture by G. Zongolopoulos and the wooden decks that lead to the entrances of the buildings (Figure 2.10).

The interior of the building is one open space developed over three split-level decks interconnected at two points by two open steel stairs with wooden treads that have been designed in such a way so as to provide minimum visual obstruction and can be seen as bridges. Metal trusses support the concrete slabs of the different levels of the building. In order to combine the structural and mechanical depth, and increase the perception of openness of the interior space, metal trusses support the concrete slabs of the different levels of the building (Figure 2.11).

Figure 2.11. *Interior spaces*

The local environmental conditions are the following:

– microclimate: suburban;

– average ambient temperature: January 10°C, July 26.5°C;

– global irradiation: 1581 kWh/m^2;

– sunshine hours: 2818 h.

The present office building is an interesting example of a well-designed building from architectural and energy point of view. The shape and orientation, as well as the interior planning of the building (it is an open space), give the possibility to exploit the local climate for energy conservation and the best use of daylight. The installation of a BEMS makes it possible to control all the systems (such as ceiling fans, night ventilation fans, etc.) installed in the building as well as to monitor energy consumption. The building is also equipped with seven split and two single package air-to-air heat pumps for cooling and heating purposes (back-up system). Rainwater is collected and used for irrigation and flushing of the double mode toilets.

The shape of the building and the location of the workstations, near the windows, allow an efficient use of the daylight. The building gets daylight from both sides by windows and from the roof by way of clerestories. These clerestories are also equipped with reflective curtains in order to distribute the daylight "deeper" into the building:

– the building is naturally lit by means of sill to ceiling windows provided on both long east- and west-facing façades and clerestories above the two stairwells;

– the east- and west-facing windows are designed with exterior fabric awnings positioned at approximately 15 cm from the wall, which provide both shading and glare control;

– the south-facing clerestory is provided with a light shelf on the south side.

Ceiling recessed downlights and task lighting compose the artificial lighting system. Artificial lighting is used only in case natural lighting is not adequate. The control of the artificial lighting system is manual on/off (Figure 2.12b).

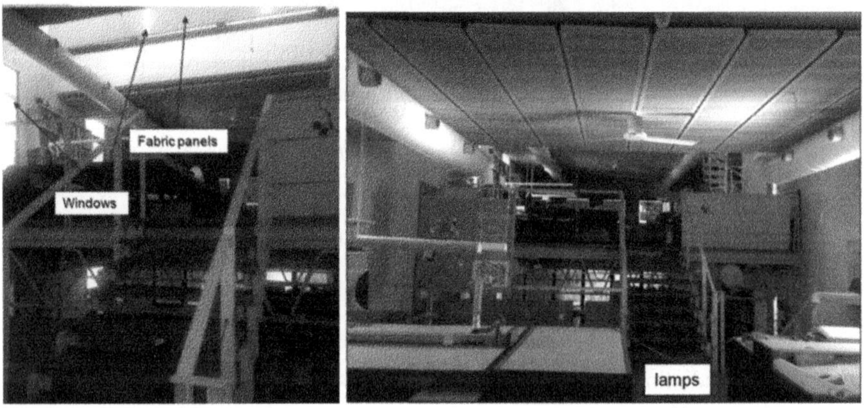

a) b)

Figure 2.12. *a) Natural and b) artificial lighting*

Another important element of the building is the shading devices. They are exterior blinds for the windows and interior blinds for the skylights. The material is white plastic cloth, and its height can be adjusted through a rotation mechanism. On the last level of the building, skylights are constructed, in the middle of the ceiling. They are covered with panels of cloth, which hang from the roof (Figure 2.12a)

Special attention has been paid to a combination of natural ventilation and passive cooling with mechanical ventilation. A central unit controls the system. In this region, the heating and cooling loads of an office-building normally cause high electricity demand. The following extra measures are taken:

– natural cooling is achieved by evaporative cooling due to the presence of vegetation and a water pond, by solar protection of the building shell and by cross ventilation;

– ventilation and cooling is achieved by cross-ventilation, when appropriate, and by automatically controlled ceiling fans, which extend the comfort range from 25°C to 29°C. When temperatures exceeds 29°C, a zoned all-air heat pump back-up air-conditioning system turns on;

– cooling is further enhanced by radiation by the mechanical night ventilation provided by two extraction fans on the roof, which by way of the lower night time exterior temperatures provide a temperature drop of about 3°C the next day;

– the thermal mass of the building is very high.

The exterior of the building is clad with fair-faced cement bricks and has increased 10 cm insulation. The U-values of the ground floor, roof and external walls are 1.41 W/m^{2}K, 0.45 W/m^{2}K, and 0.40 W/m^{2}K respectively.

Acoustics have been given special attention due to their importance in the large open interior space. Metal acoustic panels are suspended from the concrete decks of the different levels in an open configuration that allows for movement of air between the panels and ceiling.

Measurements for a full year (April 1996–March 1997) show the following results: the overall consumption (heating, cooling, lighting, night and day ventilation and equipment) is 90.1 kWh/m^2. The total consumption for heating is 35.7 kWh/m^2 and 4.7 kWh/m^2 for cooling. The energy consumption for lighting is 3.1 kWh/m^2 which on average represents 3.5% of the total energy consumed. A large amount of energy 24.7 kW/m^2 is consumed for PCs. As the building is properly designed to reduce the energy consumption for cooling, this parameter represents only the 5% of the total energy consumption. Additionally, the energy consumption for the artificial lighting system is also low (4%), proof of the right natural lighting design for the building. It is about 10 times cheaper to run this building than another of this type in Greece: in fact, office buildings have a total annual consumption (for heating and cooling) from 50 to 240 kWh/m^2/year, while the Tombazis office building consumption for heating and cooling is 41.4 kWh/m^2/year.

The energy used for heating, cooling and electricity was obtained from meter readings for nine different points in the building and from energy bills. The *total yearly energy consumption* for the monitoring period (May 2008–May 2009) was 79.7 kWh/m^2 (corrected for hours of use) with a breakdown as follows: (1) heating 29.2 kWh/m^2, (2) cooling 18.8 kWh/m^2, (3) lighting 2.7 kWh/m^2, (4) ventilation 1.2 kWh/m^2, (5) PCs 15.5 kWh/m^2, (6) other 12.3 kWh/m^2. Regarding lighting, the national average is about 30–40% of the overall building consumption, while in such a building this is up to 3.5%. The total construction cost was about 1,000,000 €.

2.13.2. *APIVITA Commercial and Industrial S.A.*

The new APIVITA building is in tune with the company's philosophy and values, symbolizing sustainability, innovation and a unique way of connecting the workplace with the natural environment. It is located in the Industrial Park of Markopoulo, Attica, Greece, in an area that was once an olive grove; this is an example of a building with NZEB principles in the design phase (Figure 2.13).

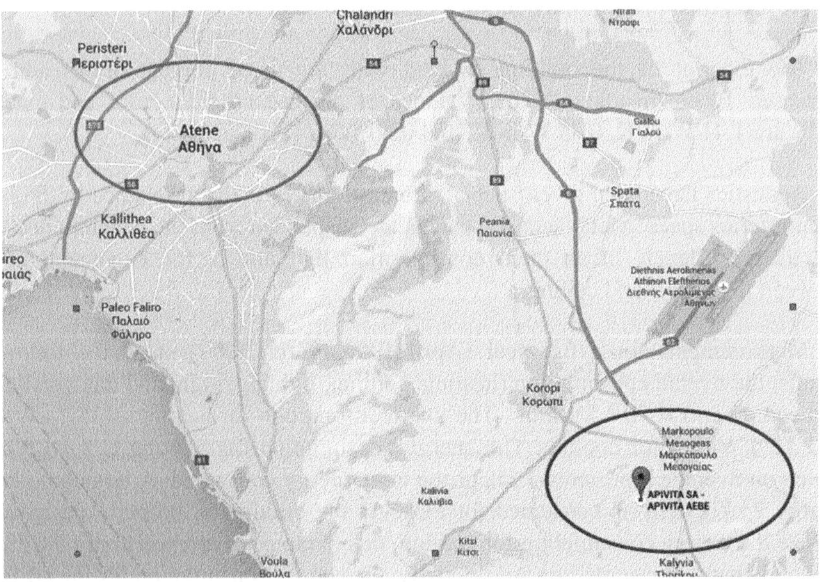

Figure 2.13. *Location of APIVITA building (red) and city center (blue)*

In terms of architectural design, the building simulates a beehive, which brings together all the company's activities, such as offices, training, museum, laboratories, cosmetics production, honey, bee pollen and royal jelly packaging, employee dining hall, SPA and events. The building covers a total surface of 6,830 m² over five levels, two underground and three aboveground, in an area of 6,170 m² (Figure 2.14).

Figure 2.14. *APIVITA interior space*

The key feature of the central atrium is an olive tree, the symbol of serenity, protection and fertility. It is known that APIVITA combines ancient Greek tradition with the rich Greek flora, focusing on the life of the bee, and creates natural, ecological, effective and holistic cosmetics. It draws inspiration from the beehive, and therefore the society of the bee. The beehive and the society of the bee lead to the main features, which are cooperativeness, dedication to a common goal, respect to one another and to the environment, the attempt to achieve high qualitative goals.

As a natural outcome of these elements, the beehive worked as an archetype for the construction of the new facilities, in terms of togetherness, allocation and use of the space, and mainly functionality, having as a dominant criteria the comfortable, day-to-day life of the employees.

As an extension of the factory, directly opposite, there is the space for interactive contact with the Greek biodiversity, a miniature of the sources of inspiration of the company: the olive trees, the working society of the bees and a labyrinth made from lavender and rosemary, which invite us to forget for a while our daily concerns, to recharge our energy levels and understand the uniqueness of the principles of the company.

The construction of the building was based on the following criteria:

– simplicity, aesthetics, and functionality;

– natural lighting and ventilation;

– respect for natural resources and local biodiversity;

– bioclimatic design;

– reduction of ecological footprint;

– thermal and visual comfort;

– recyclable, natural, and ecological Greek building materials.

To manage the natural sources, the following technologies are used:

– a system for geothermal energy exploitation;

– photovoltaics;

– a rainwater tank, where water from the building and from the environment are gathered, to be used for watering;

– biological waste cleaning and disposal of the water for watering;

– green roofs on the first and second floor;

– large plantations in the surrounding space, with aromatic, pharmaceutical Greek plants and herbs, fruit trees and species of the local flora with about 200 species and 5,000 plants in their gardens and also located on the roofs and balconies, so that the microclimate serves the biodiversity (Figure 2.15).

Figure 2.15. *Plantation in the area of APIVITA building*

The building is mainly made of glass and the energy needs are met by the geothermal energy system and PV production. Additionally, rain water from the building and from the environment is collected, filtered with a double-pass osmosis treatment and then used for watering.

Natural light infiltrates the building and in winter the solar radiation heats the air; on the contrary, in summer it helps the air ventilation due to proper openings at the top and the bottom. Moreover, interior plants improve the air quality.

The offices are situated on the first and second floor of the building, in a rectangular space with clean geometric lines. The ground floor hosts the two main production lines, namely cosmetics and honey packaging. The first floor underground hosts the packaging department, as well as the accompanying storage rooms. On the same floor we can also find the restaurant. The storage rooms for the packaging materials and the engine rooms are situated on the second floor underground.

The main goal of the architectural design was to shield the building from external environmental factors.

The shaded patio in the entrance area and the four storeys of glazing allow natural light to infiltrate the building. Thus, during winter, the solar radiation entering the building from the southern wall heats the air. During summer, it acts as a ventilation hopper due to the proper openings at the top and the bottom that facilitate air movement. The interior planting acts as a natural air purifier and maximizes the energy benefits by cooling the air.

All building materials are certified with ISO, and most of them are ecological-natural and recyclable, the spaces were constructed according to FDA standards (production, packaging, weighing, sampling) and for lighting of the offices and all internal spaces there are low consumption lights, while the outside spaces use LEDs.

In the surrounding space, there was an attempt to maintain the local flora of the area so the herbs planted outside the building are endemic, aromatic, Hippocratic plants and herbs; olive trees that existed there before were planted back and the planting was enriched with fruitful Greek and endemic trees. Further to the green roof on the first floor, a botanical garden was created with Greek aromatic, pharmaceutical and Hippocratic plants and herbs, which are used by the company in the production of its products.

About the ID process: From the very early stages, the design process of the building adopted the company's philosophy of sustainability, simplicity and functionality. All stakeholders were involved in the project from the early stages and formed a partnership. The project team worked together, adopting a holistic approach towards all major aspects: energy, environmental, comfort, economic.

2.13.3. *Stavros Niarchos Foundation Cultural Center*

The Stravros Niarchos Foundation Cultural Center (SNFCC) is located in Faliro in Greece (Figure 2.16). The SNFCC was designed by Renzo Piano and it comprises of three main parts: the National Library of Greece, the Greek National Opera, and the Stravros Niarchos Park of 170,000 m^2 aiming to restore the site's lost connections with the city and the sea (Figure 2.17). The construction of the SNFCC has created about 1,500 jobs and approximately 160 million € of annual revenue throughout the Greek economy. The SFNCC park and the roof of the opera building include Mediterranean flora: this will increase enormously in the area and new ecological systems will be introduced. The large site's green surface will cause a decrease in carbon emissions and an improvement in the air quality.

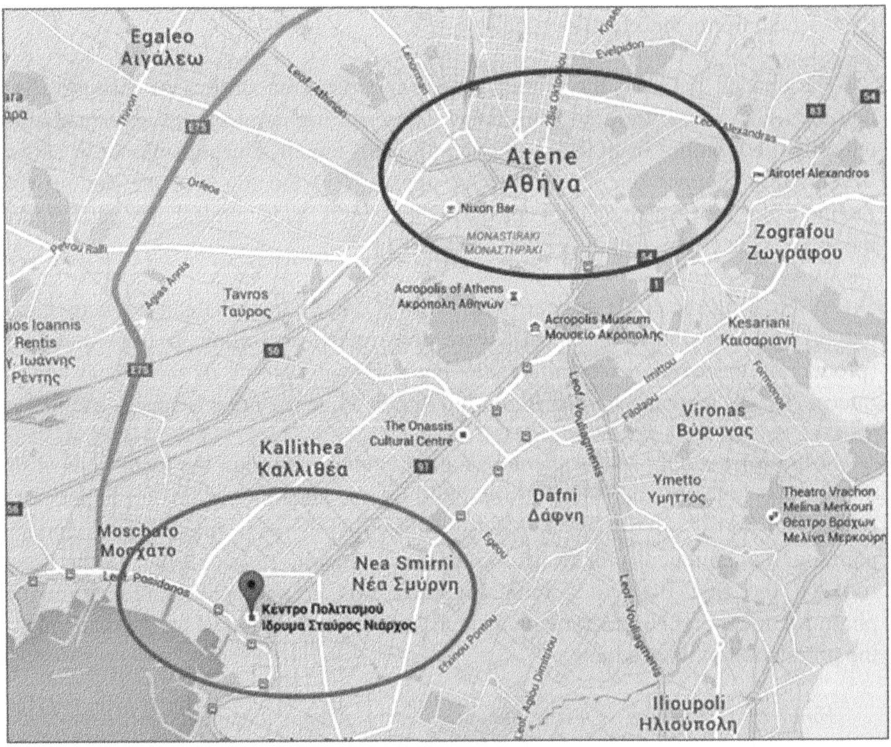

Figure 2.16. *Location of the SNFCC (in blue)*

Sustainability is one of the SNFCC's fundamental values. The creation of an environmentally friendly and sustainable infrastructure for the buildings and the park is an important goal in the design and construction of the Stavros Niarchos Foundation Cultural Center. Through environmentally innovative designs and practices, the project aims to earn at least Gold and ideally Platinum LEED certification, the first such designation in Greece and the first for a project of this scale in Europe.

Figure 2.17. *The Stavros Niarchos Foundation Cultural Center and Park © Yiorgis Yerolymbos*

As one of Athens' earliest seaports in Faliro Bay, Kallithea has always had a strong relationship with the water. To permit the view of the sea from the site, an artificial hill was created at the south (seaward) end of the site. The sloping park culminates in the cultural center building, giving it spectacular views toward the sea.

Figure 2.18. *The design of the opera*

Both the opera and the library are combined in one building, with a public space, known as the Agora, providing access and connections between the two main facilities. The opera wing is composed of two auditoria, one (450 seats) dedicated to traditional operas and ballets, the other (1,400 seats) for more experimental performances. The library is intended not only as a place for learning and preserving culture, but also as a public resource, a space where culture is truly accessible to share and enjoy (Figures 2.18 and 2.19).

Figure 2.19. *The design of the library building*

The entirely glass-walled library reading room sits on top of the building just underneath the canopy roof. A square horizontal transparent box, it will enjoy 360-degree views of Athens and the sea.

The site's visual and physical connection with water continue in the park with a new canal that will run along a north-south, main pedestrian axis, the Esplanade. The canopy roof provides essential shade and is covered with 10,000 m^2 of photovoltaic cells, enough to generate 1.5 megawatts of power for the library and opera (Figure 2.20). This field of cells should allow the building to be self-sufficient in energy terms during normal opening hours. Wherever possible, natural ventilation will be used.

The visual connection with the water will continue in the park, where it will focus on a channel to the side of the Esplanade, the main pedestrian axis of the site, in the north-south direction.

Figure 2.20. *The canopy roof*

About the ID process: before the commencement of any procurement or construction activity, all technical documents produced by the contractor (construction drawings, method statements, material submittals, etc.) are subject to the review and approval of SNFCC specialists. Such a procedure coupled with the everyday general and contractual correspondence necessitates the exchange of enormous amounts of documents within and between the three main participating entities, that is the (1) project designers, (2) employer and its representatives, and

(3) contractor and its consultants/subcontractors. The overall project design is driven by the highest environmental goals and international standards, which are summarized in the following principles: strict compliance with the environmental impact study and environmental terms approved by Greek authorities, achievement of Platinum LEED certification compliance with the highest international environmental standards. The same principles, which have already been applied in the design, are followed in everyday construction activities. The key participants in this special task are the SNFCC consultants and auditors, which include LEED specialists, third party environmental inspectors and the contractor's LEED certified supervisors. Monitoring results are documented by thorough and systematic record keeping. The status of the environmental conditions on site is constantly audited and evaluated through both periodical and ad hoc meetings recorded in monthly reports.

2.13.4. *Karelas Office Park*

The Karelas Office Park consisting of 61,574 m^2 is located southeast of Athens and follows sustainability and innovativeness concepts (Figure 2.21). The complex has a built area of 30,000 m^2 above ground and 30,000 m^2 underground. The total value of the project was 119 million euros and it was completed in 2012 (Figure 2.22). Gardens full of Mediterranean flora are spread all over the roof and there is also a path where people can walk and relax enjoying the plants and the view (Figures 2.21 and 2.24). The building was the first in Greece to be certified with a LEED GOLD certificate.

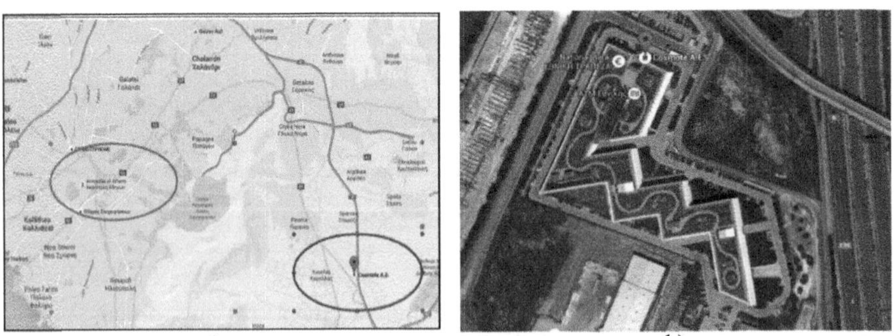

a) b)

Figure 2.21. *a) Location of Karelas Office Park; b) green rooftop of the building*

Figure 2.22. *Façade movable shading devices*

The company's aim is to achieve reductions in energy consumption whilst simultaneously satisfying the thermal comfort needs of its users. The main entrance has a glass ceiling that enables natural lighting and the walls are covered by bamboo stakes (Figure 2.23).

Figure 2.23. *Building atrium for natural day-lighting (left), atrium façade bamboo shading devices (right)*

The measures implemented within the business network include the following:

– free cooling techniques from the very beginning of the network construction;

– high efficiency rectifiers in network BTS with high load;

– modernization of the network telecom equipment;

– hybrid systems with batteries and GNs;

– use of renewable sources of energy (WT, photovoltaic systems);

– data centers operation according to the EU Code of Conduct;

– solar vacuum tubes in the main warehouse premises in Avlona;

– roof garden at the Karelas Office Park;

– building energy upgrade resulting in improved comfort conditions whilst achieving significant energy savings accomplished in the biggest call center building;

– building management systems in office buildings and data centers;

– smart metering and tele-management systems;

– energy model predicting the CO_2 footprint;

– recycling of used material (paper, etc.);

– special treatment of waste that could contaminate the environment (e.g. electronic equipment, lead batteries, etc.);

– monitoring of materials used and waste generated;

– corporate responsibility report with targets and measurements.

Figure 2.24. *Roof garden*

Following the presentations, the project participants were taken on a tour of the Karelas Office Park where some of the energy saving measures that had been implemented were presented. Figure 2.25 presents the central solar thermal system of the building for hot water needs.

Figure 2.25. *Central solar thermal system*

3

Data Analysis and Energy Modeling in Smart and Zero-energy Buildings and Communities

3.1. Energy signature for the NTL of Cyprus Institute[1]

The analysis of energy consumption data is an important instrument for a building's management: it is crucial for energy audits, to verify the savings achieved by a retrofit and to improve the estimation of expected savings (Haberl and Komor 1990). This aspect requires knowledge of the factors that influence the consumption of energy. Usually, a model of the building is necessary to estimate the consumption for any operating conditions that have to be validated.

There are two approaches for developing such a model: the simulation approach, also called forward modeling, and the system identification approach, called inverse modelling. A third option is also considered – the hybrid approach: by starting with the description of the building and calibrating the simulation program to match with the measured consumption data (Burch *et al.* 1990).

Energy model-based simulation is a computational tool that allows us to model the building as a system that is able to simulate the dynamic thermal interactions

---

Chapter written by Nikos KAMPELIS, Konstantinos GOBAKIS, Vagias VAGIAS, Denia KOLOKOTSA, Laura STANDARDI, Daniela ISIDORI, Cristina CRISTALLI, Fabio Maria MONTAGNINO, Filippo PAREDES, Pietro MURATORE, Luca VENEZIA, Marina Kyprianou DRACOU, Alaric MONTENON, Andri PYRGOU, Theoni KARLESSI and Mat SANTAMOURIS.

For a color version of all figures in this chapter, see www.iste.co.uk/kolokotsa/smartbuildings2.zip.

1 Sections 3.1 and 3.2 were written by Daniela ISIDORI.

between a building and its outdoor and indoor environments (Morbitzer 2003). This approach is useful because it is based on a deep understanding of the behavior by starting from established relationships (De Wit and Augenbroe 2002; Lam *et al.* 2004; Rizos 2007). However, conducting energy model-based simulation requires significant time, resources and technical expertise, as well as an extensive set of inputs to characterize a building (geometry, internal loads, outdoor environment, equipment, schedules), which limits their application. Usually, the customer is often not prepared to pay for such work and the soft relationship between design and analysis tools does not permit rapid energy model iteration.

The second approach, known as system identification, is based on the behavior of the building (consumption data, outdoor temperature, etc.) by assuming the structure of the model and identifying its parameters through a statistical analysis of data collected. The simplest option is to assume a steady-state model of the building and use data related to consumption, weather conditions, etc. This leads to what is known as the energy signature. The energy signature of a building represents a set of parameters that describe its energy performance by using few equations; however, a sophisticated statistical process is mandatory to make the results as reliable as possible. A very positive aspect with respect to simulation is that the system identification approach gives itself to automation. This has the advantage of requiring only data that are easy to obtain without a dedicated data acquisition system. In the literature, a classical method used for energy signature is the PRISM model (Rabl and Rialhe 1992), a regression of consumption with respect to external temperature. Other potentially important factors, such as solar radiation, wind velocity and occupancy/operating mode, particularly relevant for industrial buildings (unoccupied during holidays and weekends), are dealt with by two separate models: one for heating and one for cooling.

In the literature, numerous works have confirmed that the outdoor temperature is the central factor for buildings whose loads are dominated by heating (Rabl and Rialhe 1992). Energy signatures for air conditioning systems can have characteristics that are different with respect to heating systems, so the assumptions can be very different (Hitchin and Knight 2016).

The goal of this work is to calculate the energy signature of the Novel Technology Lab in Athalassa campus of Nicosia, to obtain the overall estimation of the consumption by starting from common measurements that are easily collected in a smart building, such as external and internal temperature (Figure 3.1). In this way, one can bypass the simulation model and directly obtain a load estimation that can be used to tune and optimize renewable production. A TRNSYS model of NTLab, together with Fresnel technology, has already been developed and explained in Kolokotsa *et al.* (2016b). Then, once the loads estimation of the building are obtained, it is possible to insert in the TRNSYS model the numerical equations obtained with a statistical analysis of the collected data.

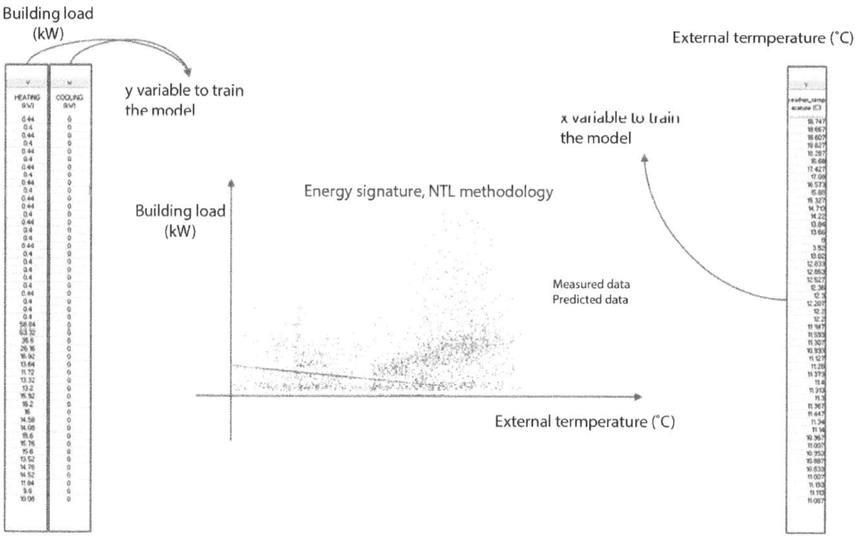

Figure 3.1. *Energy signature, conceptual schema*

Estimation of energy performance indexes, like the heating/cooling curve or the energy signature, requires robust regression models of the heating or cooling losses on the outdoor temperature.

The model gives the overall heating/cooling loads with respect to external temperature, and it may be used to estimate the energy consumption for datasets of the outdoor temperature coming from different time–space locations with an acceptable error for a preliminary analysis.

Since the aim of the task is to understand the sizing and positioning as well as the effective operation of renewables for smart grids, the solar thermal Fresnel collectors of CyI is exploited as part of a smart grid. For this reason, improvements via advanced control and data mining techniques for energy predictions have been explored.

3.2. Athalassa Campus and the NTL building

The Fresnel technology began to be part of a potential grid, composed by many buildings of Athalassa campus (adjacent to French Cypriot School; see Figure 3.2) within Aglantzia Municipality in Nicosia, where the Cyprus Institute is located. The

Cyprus Institute increasingly enclosed additional space for use, from non-used buildings at the Campus to expansions that required laboratory and office space.

In 2016, the main facilities within the Campus were (Rabl and Rialhe 1992):

– The Guy Ourisson Building (2007): the first renovated building of CyI that includes a combination of offices (for faculty, researchers and administration), seminar and conference rooms as well as auxiliary common areas, the cafeteria, office space for students, a students' common area in the interconnected Graduate School building and an external theatre.

– The Novel Technologies Laboratory (NTL) (2014): designed as a "green building", using specific standards that permit it to be a near zero-energy building. The laboratory hosts offices and laboratory teams from the Energy, Environment and Water Research Centre and is equipped with the latest technology equipment for scientific research.

– The Cy-Tera, CyI's High Performance Computing Facility (2012): one of most powerful supercomputers within the region used for scientific research. The building includes office space for the researchers and the technical team.

– STARLab Laboratory: infrastructure for the scientific research, documentation, and preservation of the cultural heritage of Cyprus. The facility includes a fixed and a mobile laboratory.

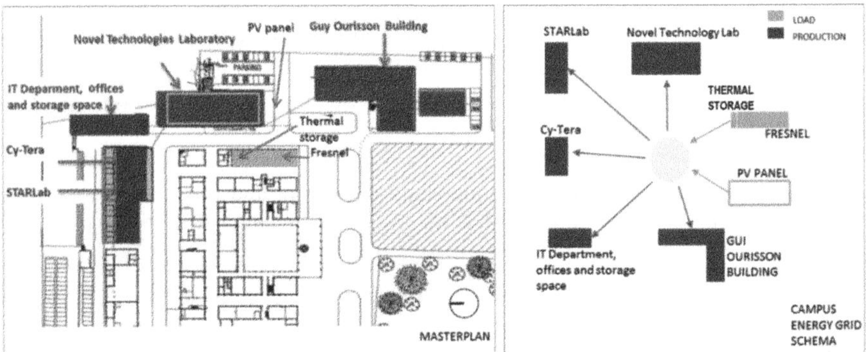

Figure 3.2. *Athalassa campus and energy grid scheme*

A preliminary focus on the campus started by considering the buildings as "load" (consumption) or "production" nodes that compose the smart grid. Smart grids

(Amin and Wollenberg 2005) provide the integration of traditional and renewable energy resources, with new elements in a distributed, open and self-managed way. This requires innovative models for energy infrastructure and the self-management of the power grid (Kephart and Chess 2003) to ensure the scalability of smart grids and allow the management of autonomous operations throughout the power network.

In this context, the existing integrations between two smart grid nodes of the Athalassa Campus "potential" grid, namely the Novel Technologies Laboratory (NTL) and the linear Fresnel reflector (LFR) with thermal storage were considered. The LFR system consists of a Fresnel solar collector, an oil buffer storage, a heat exchanger to generate hot water and a single stage chiller to be integrated with the existing HVAC (heating, ventilation, air conditioning) units. The solar collector has been designed by Idea Srl and addresses the possibility to install PV panels on the back of the primary optics, and to produce electricity when the direct solar radiation is poor.

The aim of the linear Fresnel reflector is to explore the opportunities offered by solar poly-generation for near zero-energy buildings and settlements, to maximize the utilization of the collected solar energy to the largest possible extent. The research activities during the first years of Smart GEMS have been devoted to the study of the main issues emerging in this kind of integrated system and, specifically, on tuning, liability, automation, building integration and hybridization with the in-place air conditioning systems (Kolokotsa *et al.* 2016b).

The NTL with a linear Fresnel reflector and combined HVAC systems has been assessed in terms of the minimization of consumption and full exploitation of solar plant integration. In addition, dynamic modeling and the evaluation of concentrated solar power (CSP) technology and in specific, the IDEA FRESCO system was investigated as a solution for integration in nearly zero-energy buildings (NZEBs) and smart grids (Kolokotsa *et al.* 2016b).

A model of the Fresnel system was developed and exploited to fully understand the operational phase and coupling possibilities of such systems (Figure 3.3). In such a context, the goal is to give an overall estimation of energy consumption by using a data driven model obtained with the data collected in Novel Technology Laboratory (Figure 3.4).

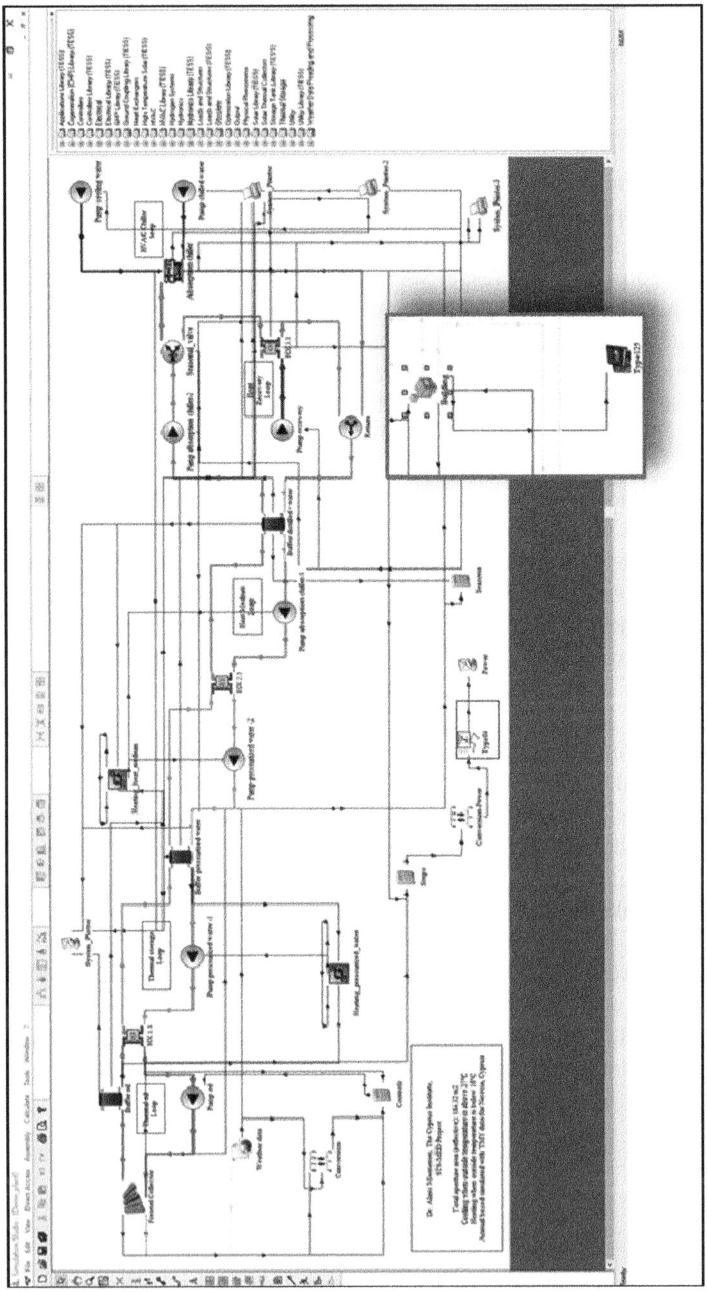

Figure 3.3. *Fresnel and thermal loop TRNSYS model*

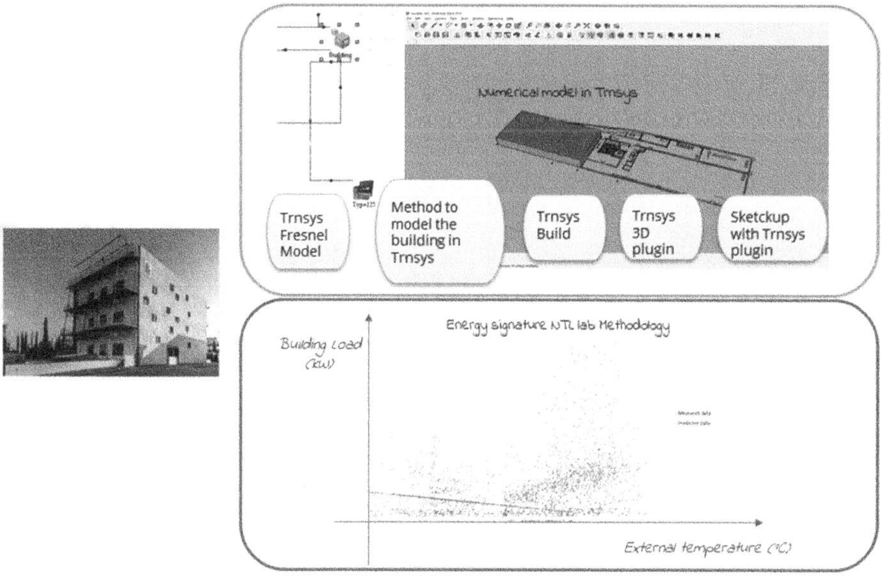

Figure 3.4. *Simulation model versus data driven model of the NTL*

3.2.1. *Methodology*

The data analysis general methodology is composed of several fundamental steps that start by defining the problem. After the phases of data exploration and data preparation, data analysis is conducted through testing several algorithms, related to supervised and unsupervised techniques (Figure 3.5).

Data preparation has been obtained through Microsoft Excel and the Labview software environment while algorithms used for data analysis have been developed in the Python programming language.

In this case, regression analysis is employed in order to model its relationship among power load, external temperature and other collected data.

Then, the outcome (Y) based on values of a set of predictor variables (X_i) has been predicted by using the regressive method.

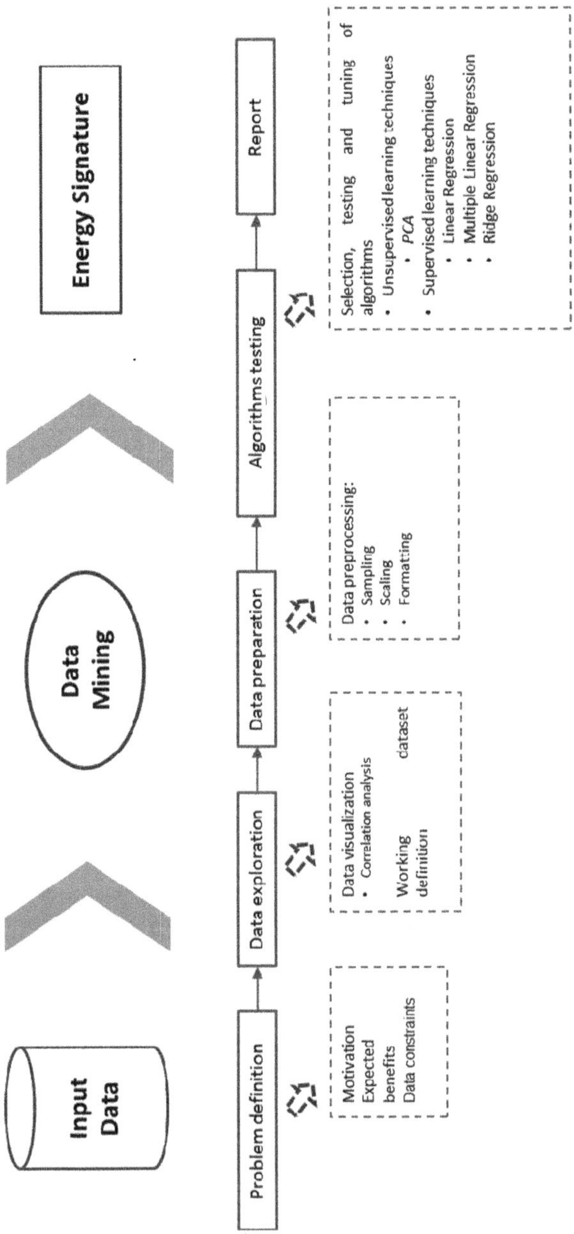

Figure 3.5. *Flow chart*

Regression analysis is widely used for the prediction and forecasting of a quantity and helps us understand how the value of the dependent variable changes when any one of the independent variables ("predictors") is varied, while the other predictors are held fixed. The regression function is a function of the independent variables for the estimation target.

These methods allow one to evaluate the impact of multiple variables in the same model. In the literature, there are various types of regression analysis.

The starting point is a linear regression model in order to add each time a new (X_i) variable so as to better fit the Y outcome. Since the first results with a linear model are not as good as expected, we introduce non-linear regression with the ridge technique to create the most reliable model possible for a preliminary study of the building consumption.

3.2.2. *Description of the Novel Technology case study*

As previously mentioned, the Novel Technologies Laboratory mainly consists of offices and laboratories on four levels; the basement, ground floor, first floor and second floor. The basement contains laboratories, a WC and two offices. The ground floor is composed of the main lobby/entrance, two laboratories, six offices, a storage room, a small kitchen and a WC. The first floor has a lecture hall, five offices, a conference room, a small kitchen and a WC. The second floor contains a laboratory, five offices, a kitchen and a WC.

The Novel Technologies Laboratory operates 8 hours/day, from Monday to Friday and for about 240 days/year.

The demand of energy for the NTL has a distribution that is typical in Mediterranean countries. Figure 3.6 shows the power demand required over the course of a year: the red line indicates the average basic load demand (general appliances elevator, ventilation, hot water) and the blue line shows the artificial lighting, while remaining electricity consumption, in cold and warm months, refers to the load demand for heating and cooling.

ETL (extract, transform, load) procedures started from "load demand" data and "meteo station" data that has been collected in two separates files with a different sample rate (Figure 3.7).

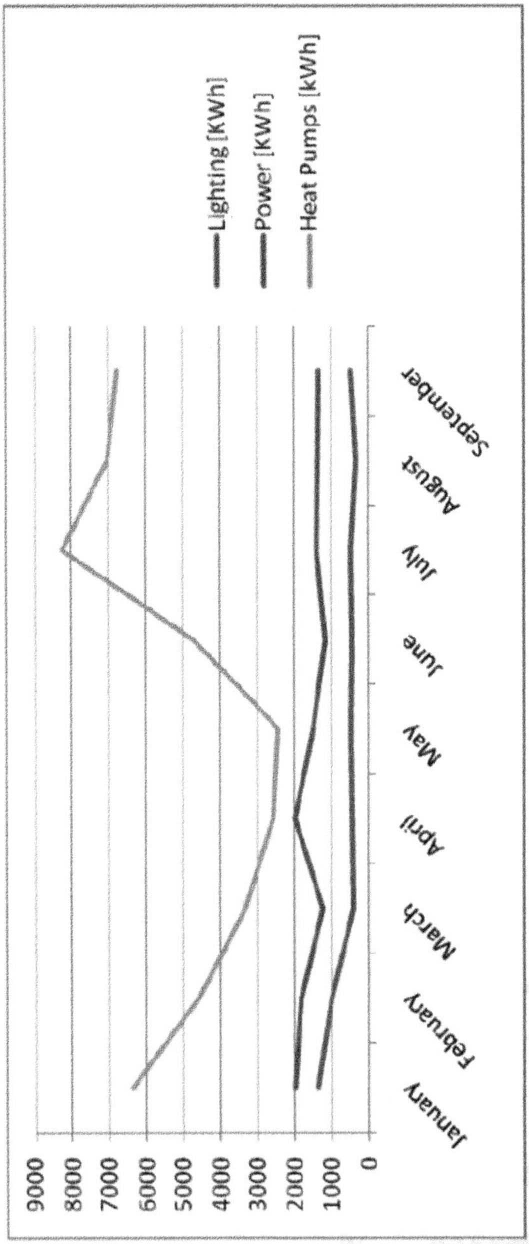

Figure 3.6. *NTL power demand*

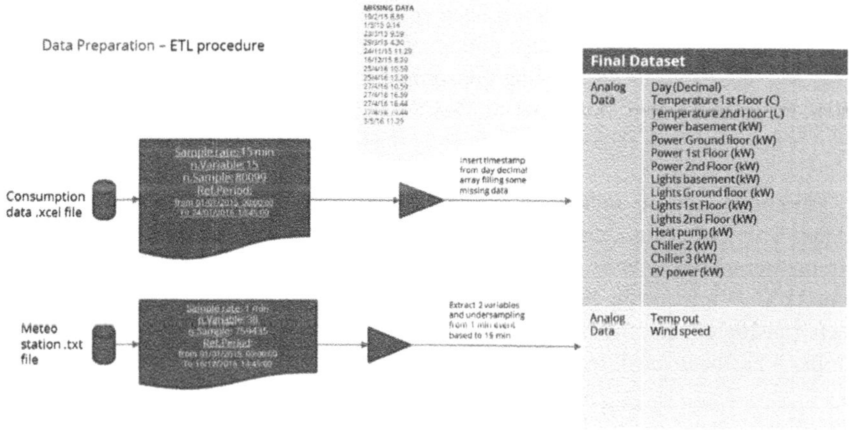

Figure 3.7. *ETL procedure*

Sub-sampling procedures have been developed to be scalable and programmed using the Labview platform (Figure 3.8).

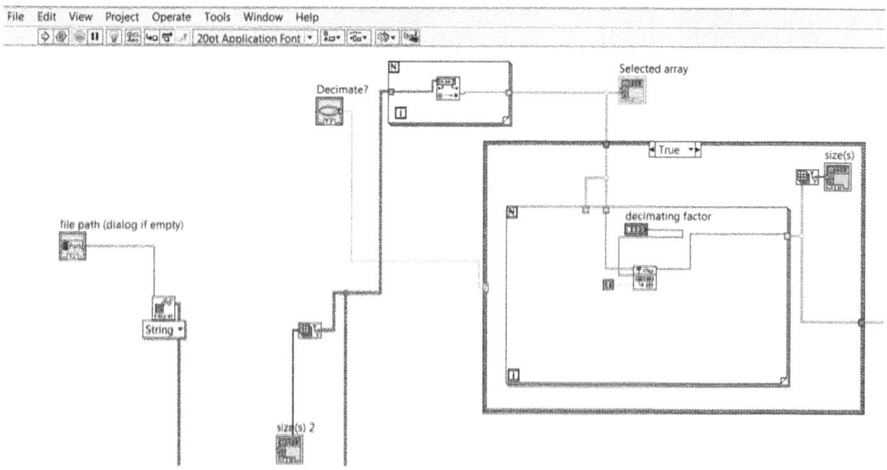

Figure 3.8. *Re-sampling algorithm*

Another important aspect to consider is the correct timestamp in order to understand the real consumption levels in operative settings. For this reason, several conditions related to working time, the weekend and national holidays that select only the most relevant data and exclude all data collected when the building is not fully operative have been added.

Moreover, the original data has been split in two datasets, in order to obtain two kinds of energy signature models: one is related to the heating consumption, and the other one is related to the cooling consumption. For this reason, both for 2015 and 2016, the period from January 01 to May 01 and the period after November 14 have been considered with regards to the heating dataset and the cooling dataset.

Another condition, related maximum or minimum external temperature, has been added. This is because, for a fully operative condition, during heating period, the external temperature has to be lower than 25°C and during summer has to be higher than 25°C. While for the heating dataset only the consumption of the Heat Pump has been considered, for the cooling period the sum among Heat Pump, Chiller 2 and Chiller 3 has been used. In Table 3.1, all the added conditions are explained.

Select working time	Dataset from 9:00 am to 05:00 pm
Delete weekend day	Delete all samples collected during Saturday and Sunday
Delete national holiday	Delete all samples collected during national holidays and expected vacation days
Heating variable HEATING (kW)	– Time t > 01/01/2015 and t < 01/05/2015 or t > 14/11/2015 – Weather Temperature T < 25°C HEATING (kW) = Heat Pump (kW)
Cooling variable COOLING (kW)	– Time 01/01/2015 < t < 01/05/2015 or t > 14/11/2015 – Weather Temperature T > 25°C COOLING (kW) = Heat Pump (kW) + Chiller 2 (kW) + Chiller 3 (kW)

Table 3.1. *Dataset filters*

The final dataset consists of 27 columns, corresponding to the observed variables, and a number of rows corresponding to the various samples collected, with 15 minutes, between January 1, 2015 and December 31, 2016 (Figure 3.9).

Other two additional variables are added. One, called "*orari*", is obtained by extracting a single hour from the timestamp. The other one, called "*deltaT*", is obtained by calculating the difference between the average of the internal and external temperature.

Finally, the dataset has been split again in order to obtain the training dataset, which is completely different from the test dataset. The training set is a set of data used to discover potentially predictive relationships and the test set is a set of data used to assess the strength and utility of a predictive relationship. The initial discovery of relationships is usually done with a training set, while a test set and a validation set are used for evaluating whether or not the discovered relationships hold. In this case, the whole of 2015 is the training set and 2016 is the test set (Figure 3.10).

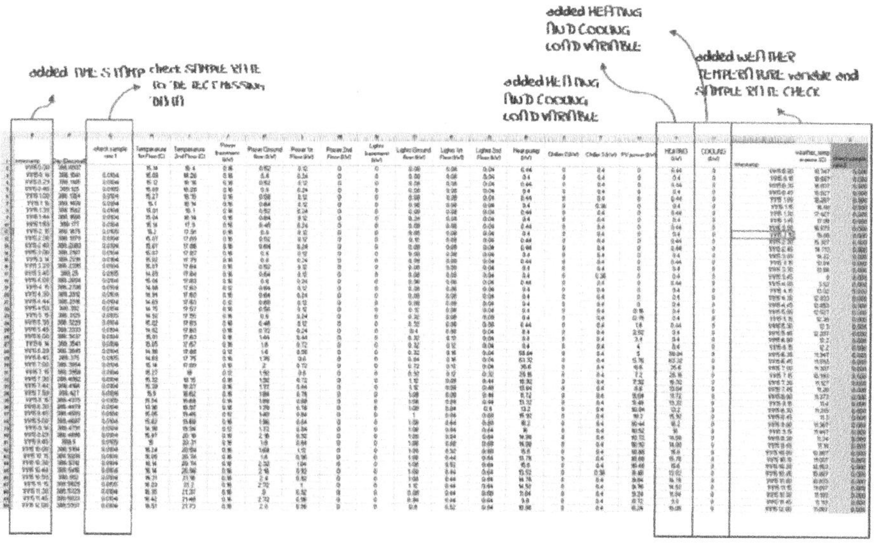

Figure 3.9. *Selected dataset*

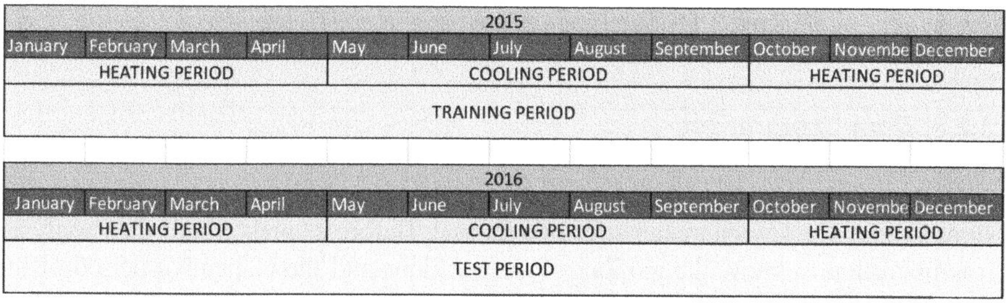

Figure 3.10. *Training and test dataset*

The energy signature (heating and cooling consumption versus external temperature) of the selected dataset with respect to the original dataset is shown in Figure 3.11.

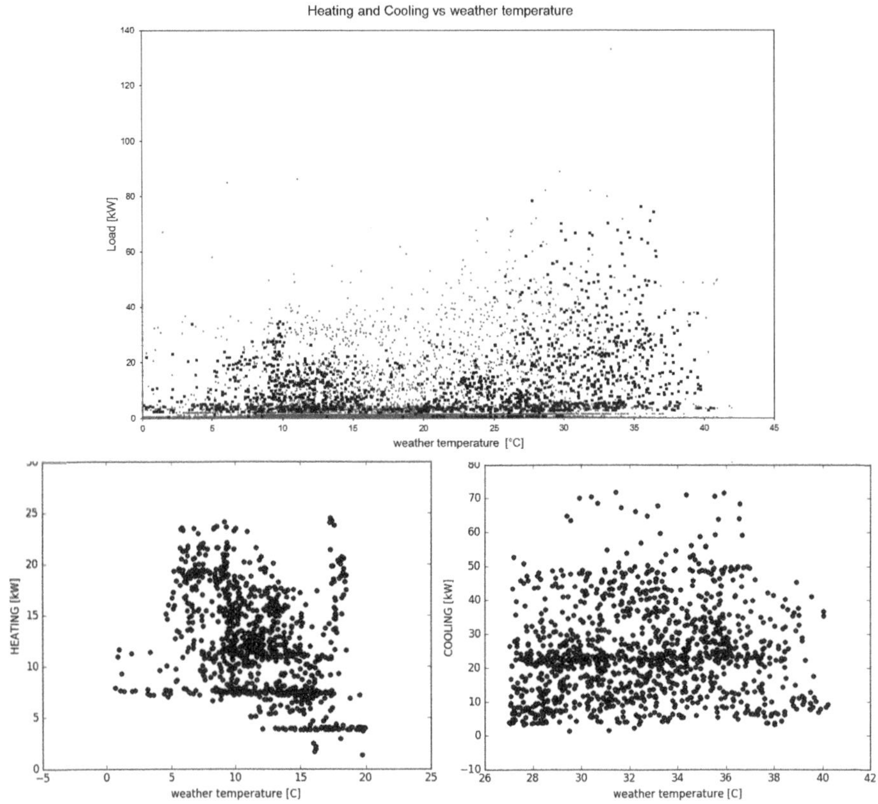

Figure 3.11. *Selected dataset after the ETL procedure*

3.2.3. *Data exploration*

Data exploration is the first step in data analysis and usually involves summarizing the key characteristics of a dataset. Before a "prescribed" data analysis, it is important to know the number of observations and the respective period of the dataset, variables included, the missing data, average values and the general hypotheses the data is likely to support.

From this step, one can recognize variables expected to lead to interesting observations. For example, by displaying the mean external temperature variable we can simply check if the two clusters for heating and cooling are appropriated and coherent with thermal conditions. In Table 3.2 and Table 3.3, the information related to the number of samples, mean value, standard deviation, min, 25%, 50%, 75% and maximum value are shown for each variable.

	Temperature 1st Floor (C)	Temperature 2nd Floor (C)	Heat pump (kW)	Chiller 2 (kW)	Chiller 3 (kW)	PV power (kW)	weather_temperature	orari	weekdays	deltaT	heating
count	2396	2396	2396	2396	2396	2396	2396	2396	2396	2395	2396
mean	17.16255843	19.0924207	10.52388982	0.180033389	0.400617696	4.181619366	11.78800793	12.62	2.071368948	-6.3394815	10.5238898
std	1.51954765	1.846996587	6.967868963	0.200660285	0.01267361	4.191526521	3.814949122	2.36	1.392913156	3.8761793	6.96786896
min	11.73	12.51	1.04	0	0.36	0	0.593	9	0	-20.723	1.04
25%	16.31	18.15	4.04	0	0.4	0.12	9.4	11	1	-8.942	4.04
50%	17.11	19.41	7.84	0	0.4	2.84	11.78	13	2	-6.3415	7.84
75%	17.82	20.2	15.32	0.4	0.4	7.97	14.1585	15	3	-3.8895	15.32
max	23.01	23.95	36.16	0.44	0.44	13.68	19.307	17	4	3.66	36.16

Table 3.2. *Data exploration for a heating dataset*

	Temperature 1st Floor (C)	Temperature 2nd Floor (C)	Heat pump (kW)	Chiller 2 (kW)	Chiller 3 (kW)	PV power (kW)	weather_temperature	orari	weekdays	deltaT	cooling
count	2362.000	2362.000	2362.000	2362.000	2362.000	2362.000	2362.000	2362	2362.000	2362.000	2362.000
mean	25.0292083	25.8142337	10.24733277	4.840321761	5.129077053	6.61703641	32.31518967	12.74	2.033022862	6.8934687	20.2167316
std	2.085831047	2.445781115	16.49763977	10.39229874	10.52949692	3.783842645	3.245778846	2.25	1.420999436	3.2185138	16.5229615
min	20	21.87	0.32	0.36	0.36	0	27.013	9	0	-2.297	1.12
25%	23.58	23.91	0.44	0.44	0.44	2.89	29.5485	11	1	4.61775	6.97
50%	25	24.94	0.44	0.44	0.44	7.5	32.307	13	2	6.56	16.02
75%	26.59	27.37	14.42	1.35	2.79	10.2	34.6115	15	3	9.21925	27.16
max	30.49	30.63	87.24	85.2	69.96	12.04	40.727	17	4	17.142	88.12

Table 3.3. *Data exploration for a cooling dataset*

3.2.4. *Correlation matrix*

Pearson's correlation coefficient between two variables is defined as the covariance of the two variables divided by the product of their standard deviations.

Correlation is computed into what is known as the correlation coefficient, which ranges between -1 and +1. A perfect positive correlation (a correlation coefficient of +1) implies that as one variable moves, either up or down, the other variable will move in lockstep, in the same direction. A perfect negative correlation means that if a variable changes in a direction, then the variable that is perfectly negatively correlated will change in the opposite direction. If the correlation is 0, there is no correlation; they are completely random.

In Tables 3.4 and 3.5, the correlation matrix for both datasets are shown. One can observe several levels of correlation.

	Temperature 1	Temperature 2	Power baseme	Power Ground	Power 1st Floc	Power 2nd Flo	Lights baseme	Lights Ground	Lights 1st Floo	Lights 2nd Floc	Heat pump (kV	Chiller 2 (kW)	Chiller 3 (kW)	PV power (kW)	weather_temp	orari	deltaT	heating
Temperature 1st Floor (C)	1.00	0.73	0.09	0.22	0.25	0.11	0.12	0.20	0.15	0.19	-0.05	-0.13	0.19	0.00	0.16	0.13	-0.21	-0.05
Temperature 2nd Floor (C)	0.73	1.00	0.14	0.32	0.40	0.16	0.23	0.28	0.21	0.41	0.28	-0.29	0.13	0.06	0.15	0.06	-0.23	0.28
Power basement (kW)	0.09	0.14	1.00	0.23	0.09	0.54	0.46	0.05	0.05	0.29	0.25	0.20	-0.02	0.05	0.13	-0.02	0.07	0.25
Power Ground floor (kW)	0.22	0.32	0.23	1.00	0.43	0.18	0.23	0.20	0.09	0.56	0.54	-0.07	-0.02	0.18	-0.03	-0.28	-0.15	0.54
Power 1st Floor (kW)	0.25	0.40	0.09	0.43	1.00	0.03	0.14	0.14	0.14	0.40	0.48	-0.08	0.00	0.19	0.00	-0.28	-0.14	0.48
Power 2nd Floor (kW)	0.11	0.16	0.54	0.18	0.03	1.00	0.55	0.23	0.03	0.30	0.10	0.57	-0.06	-0.05	0.39	0.01	0.33	0.10
Lights basement (kW)	0.12	0.23	0.46	0.23	0.14	0.55	1.00	0.28	0.13	0.55	0.20	0.41	-0.03	-0.01	0.21	-0.06	0.13	0.20
Lights Ground floor (kW)	0.20	0.28	0.05	0.20	0.14	0.23	0.28	1.00	0.52	0.31	0.19	0.28	-0.14	-0.46	0.02	0.37	-0.08	0.19
Lights 1st Floor (kW)	0.15	0.21	0.05	0.09	0.14	0.03	0.13	0.52	1.00	0.16	0.15	-0.03	-0.09	-0.32	-0.07	0.34	-0.15	0.15
Lights 2nd Floor (kW)	0.19	0.41	0.29	0.56	0.40	0.30	0.55	0.31	0.16	1.00	0.43	0.16	-0.04	0.08	0.06	-0.21	-0.07	0.43
Heat pump (kW)	-0.05	0.28	0.25	0.54	0.48	0.10	0.20	0.19	0.15	0.43	1.00	-0.10	-0.07	0.01	-0.12	-0.13	-0.18	1.00
Chiller 2 (kW)	-0.13	-0.29	0.20	-0.07	-0.08	0.57	0.41	0.28	-0.03	0.16	-0.10	1.00	-0.11	-0.16	0.21	0.00	0.30	-0.10
Chiller 3 (kW)	0.19	0.13	-0.02	-0.02	0.00	-0.06	-0.03	-0.14	-0.09	-0.04	-0.07	-0.11	1.00	0.09	-0.03	0.05	-0.10	-0.07
PV power (kW)	0.00	0.06	0.05	0.18	0.19	-0.05	-0.01	-0.46	-0.32	0.08	0.01	-0.16	0.09	1.00	0.11	-0.74	0.09	0.01
weather_temperature	0.16	0.15	0.13	-0.03	0.00	0.39	0.21	0.02	-0.07	0.06	-0.12	0.21	-0.03	0.11	1.00	-0.11	0.92	-0.12
orari	0.13	0.06	-0.02	-0.28	-0.28	0.01	-0.06	0.37	0.34	-0.21	-0.13	0.00	0.05	-0.74	-0.11	1.00	-0.15	-0.13
deltaT	-0.21	-0.23	0.07	-0.15	-0.14	0.33	0.13	-0.08	-0.15	-0.07	-0.18	0.30	-0.10	0.09	0.92	-0.15	1.00	-0.18
heating	-0.05	0.28	0.25	0.54	0.48	0.10	0.20	0.19	0.15	0.43	1.00	-0.10	-0.07	0.01	-0.12	-0.13	-0.18	1.00

Table 3.4. *Correlation matrix for a heating dataset. For a color version of this table, see www.iste.co.uk/kolokotsa/smartbuildings2.zip*

Regarding the heating dataset, the results obtained are:

– Average correlated variables: Power Ground floor, Power 1st floor, and Light 2nd floor.

– Weakly correlated variables: weather temperature, orari, delta T, Light basement, Light Ground floor, Light 1st floor, Chiller 2, Chiller 3, PV power.

Regarding the cooling dataset, the results obtained are:

– Average correlated variables: Power 1st floor, Power 2nd floor, Light 1st floor and Light 2nd floor.

– Weakly correlated variables: orari, delta T, Light basement, Light Ground floor, Light 1st floor, Chiller 2, Chiller 3, PV power.

COOLING	Temperature 1	Temperature 2	Power baseme	Power Ground	Power 1st Floc	Power 2nd Flo	Lights basemei	Lights Ground	Lights 1st Floo	Lights 2nd Floc	Heat pump (kV	Chiller 2 (kW)	Chiller 3 (kW)	PV power (kW)	weather_temp	orari	deltaT	cooling
Temperature 1st Floor (C)	1.00	0.81	-0.07	-0.04	-0.56	-0.06	-0.13	-0.47	-0.35	-0.59	-0.04	-0.22	-0.19	-0.09	0.37	0.16	-0.26	-0.29
Temperature 2nd Floor (C)	0.81	1.00	-0.12	-0.08	-0.52	-0.08	-0.15	-0.50	-0.33	-0.66	-0.13	-0.21	-0.22	-0.10	0.29	0.16	-0.35	-0.40
Power basement (kW)	-0.07	-0.12	1.00	0.01	0.21	0.18	0.92	0.14	0.01	0.20	-0.04	0.21	0.21	0.10	0.23	-0.02	0.30	0.22
Power Ground floor (kW)	-0.04	-0.08	0.01	1.00	0.07	0.07	0.01	0.07	0.02	0.07	0.09	0.03	0.05	0.06	0.00	-0.08	0.05	0.14
Power 1st Floor (kW)	-0.56	-0.52	0.21	0.07	1.00	0.22	0.23	0.48	0.38	0.62	0.26	0.20	0.29	0.30	-0.04	-0.31	0.33	0.57
Power 2nd Floor (kW)	-0.06	-0.08	0.18	0.07	0.22	1.00	0.16	0.22	-0.03	0.08	0.39	0.05	0.18	0.06	0.18	-0.07	0.23	0.54
Lights basement (kW)	-0.13	-0.15	0.92	0.01	0.23	0.16	1.00	0.17	0.09	0.26	-0.03	0.20	0.16	0.10	0.17	-0.04	0.27	0.20
Lights Ground floor (kW)	-0.47	-0.50	0.14	0.07	0.48	0.22	0.17	1.00	0.26	0.56	0.09	0.16	0.22	0.18	-0.14	-0.18	0.20	0.33
Lights 1st Floor (kW)	-0.35	-0.33	0.01	0.02	0.38	-0.03	0.09	0.26	1.00	0.37	0.09	0.01	0.08	0.03	-0.21	-0.08	0.03	0.15
Lights 2nd Floor (kW)	-0.59	-0.66	0.20	0.07	0.62	0.08	0.26	0.56	0.37	1.00	0.19	0.21	0.21	0.28	-0.19	-0.29	0.25	0.45
Heat pump (kW)	-0.04	-0.13	-0.04	0.09	0.26	0.39	-0.03	0.09	0.09	0.19	1.00	-0.25	-0.26	0.11	-0.03	-0.15	0.03	0.67
Chiller 2 (kW)	-0.22	-0.21	0.21	0.03	0.20	0.05	0.20	0.16	0.01	0.21	-0.25	1.00	-0.19	0.16	0.05	-0.13	0.20	0.26
Chiller 3 (kW)	-0.19	-0.22	0.21	0.05	0.29	0.18	0.16	0.22	0.08	0.21	-0.26	-0.19	1.00	0.14	0.13	-0.14	0.28	0.26
PV power (kW)	-0.09	-0.10	0.10	0.06	0.30	0.06	0.10	0.18	0.03	0.28	0.11	0.16	0.14	1.00	-0.07	-0.87	-0.01	0.30
weather_temperature	0.37	0.29	0.23	0.00	-0.04	0.18	0.17	-0.14	-0.21	-0.19	-0.03	0.05	0.13	-0.07	1.00	0.26	0.78	0.09
orari	0.16	0.16	-0.02	-0.08	-0.31	-0.07	-0.04	-0.18	-0.08	-0.29	-0.15	-0.13	-0.14	-0.87	0.26	1.00	0.15	-0.32
deltaT	-0.26	-0.35	0.30	0.05	0.33	0.23	0.27	0.20	0.03	0.25	0.03	0.20	0.28	-0.01	0.78	0.15	1.00	0.33
cooling	-0.29	-0.40	0.22	0.14	0.57	0.54	0.20	0.33	0.15	0.45	0.67	0.26	0.26	0.30	0.09	-0.32	0.33	1.00

Table 3.5. *Correlation matrix for a cooling dataset. For a color version of this table, see www.iste.co.uk/kolokotsa/smartbuildings2.zip*

Obviously, both heating and cooling variables have a very strong correlation with the heat pump (in the case of the heating variable, it is exactly the same variable). Instead, the cooling dataset does not have much in correlation with the external temperature and for this reason a multivariable regression model is necessary to obtain more reliable results.

3.2.5. *Regression model*

The techniques for implementing regression analysis are: parametric, such as linear regression and ordinary least squares regression, and non-parametric regression.

Regression models are often fitted using the least squares method, but they may also be fitted by minimizing the "lack of fit" in some other norm (as with least absolute deviations regression), or by minimizing a penalized version of the least squares loss function, as in ridge regression (L2-norm penalty) and Lasso (L1-norm penalty) (Crawley *et al.* 1999).

In this case, as previously described, two arrays, *x_train* and *y_train* have been created: *x_train* is related to the independent variable, and *y_train* is instead the dependent variable.

The result of regression analysis methods depends on the procedure of the data generating process, and how it relates to the regression approach being used.

There are three different approaches to evaluate the quality of predictions of a model:

– *Estimator score method*: estimators have a score method providing a default evaluation criterion for the problem they are designed to solve.

– *Scoring parameter*: model-evaluation tools using cross-validation rely on an internal scoring strategy.

– *Metric functions*: the metrics module implements functions assessing prediction error for specific purposes. These metrics are detailed in sections on classification metrics, multilabel ranking metrics, regression metrics and clustering metrics.

In this work, regression metrics are used by considering several losses, the score and utility functions to measure regression performance. Some of those have been enhanced to handle the multi-output case: mean squared error, mean absolute error, explained variance score and r2_score.

The r2_score and explained variance score accept an additional value "variance weighted" for the multi-output parameter. This option brings to a weighting of each individual score by the variance of the corresponding target variable. This setting quantifies the globally captured unscaled variance. If the target variables are of different scale, then this score puts more importance on explaining the higher variance variables.

3.2.5.1. *Explained variance score*

The explained variance score computes the explained variance regression score. If $y_{i\,pred}$ is the predicted value and $y_{i\,train}$ is the corresponding true value, and Var is Variance, the square of the standard deviation, then the explained variance is estimated to be:

$$Explained\ variance\ (y_{train}, y_{pred}) = 1 - \frac{Var\,\{y_{i\,train} - y_{i\,pred}\}}{Var\,\{y_{i\,train}\}}$$

and the best possible score is 1.0.

3.2.5.2. *Mean absolute percentage error*

Another metric normally used to evaluate prediction accuracy of the regression model is the mean absolute percentage error (MAPE), defined by the formula:

$$M = \frac{100}{n} \sum_{t=1}^{n} \left| \frac{A_t - F_t}{A_t} \right|$$

where A_t is the actual value and F_t is the forecasted value.

The difference between A_t and F_t is divided by the Actual Value A_t again (cannot be used if there are zero values). The absolute value in this calculation is summed for every forecasted point in time and divided by the number of fitted points n.

3.2.5.3. *Mean absolute error*

The mean absolute error function corresponds to the expected value of the absolute error loss or L1-norm loss.

If $y_{i\,pred}$ is the predicted value of the ith sample, and $y_{i\,train}$ is the corresponding true value, then the mean absolute error (MAE) estimated over n is defined as:

$$MAE\ (y_{train}, y_{pred}) = \frac{1}{n} \sum_{i=1}^{n-1} \left| y_{i\,train} - y_{i\,pred} \right|$$

3.2.5.4. *Mean squared error and root mean squared error*

The mean squared error function and root squared error compute mean square error and its root, a metric corresponding to the expected value of the squared (quadratic) error loss or root error loss.

If $y_{i\,pred}$ is the predicted value of the ith sample, and $y_{i\,train}$ is the corresponding true value, then the mean squared error (MSE) and the root mean squared error (RMSE) is estimated over n defined as:

$$MSE\ (y_{train}, y_{pred}) = \frac{1}{n} \sum_{i=1}^{n-1} \left(y_{i\,train} - y_{i\,pred} \right)^2$$

$$RMSE\ (y_{train}, y_{pred}) = \frac{1}{n} \sum_{i=1}^{n-1} \sqrt{\left(y_{i\,train} - y_{i\,pred} \right)^2}$$

3.2.5.5. *Median absolute error*

The median absolute error is calculated by taking the median of all absolute differences between the target and the prediction.

If $y_{i\,pred}$ is the predicted value of the ith sample, and $y_{i\,train}$ is the corresponding true value, then the median absolute error (MedAE) estimated over n is defined as:

$$MedAE\ (y_{train}, y_{pred}) = median(|y_{1\,train} - y_{1\,pred}|, \dots, |y_{n\,train} - y_{n\,pred}|)$$

3.2.5.6. *R^2 score function*

The R^2 score function computes R^2, the coefficient of determination. It provides an estimation of how fine future samples are likely to be predicted by the model. It can also be interpreted as the amount of variation of the dependent variable explained by the regression equation (Fan *et al.* 2008). The best possible score is 1.0. A constant model that always predicts the expected value of y, disregarding the input features, would get an R^2 score of 0.0.

If $y_{i\,pred}$ is the predicted value of the ith sample, and $y_{i\,train}$ is the corresponding true value, then the R^2 score estimated over n is defined as:

$$R^2\ (y_{train}, y_{pred}) = 1 - \frac{\sum_{i=1}^{n-1} (y_{i\,train} - y_{i\,pred})^2}{\sum_{i=1}^{n-1} (y_{i\,train} - \bar{y})^2}$$

where $\bar{y} = \frac{1}{n} \sum_{i=0}^{n-1} y_{i\,train}$

3.2.5.7. *Linear regression model*

The simple linear regression model to express the dependence of the expected response μ_i on the predictor x_i can be written as:

$$Y = \beta_0 + \beta_i x_i$$

where β_0 is called the constant or intercept, and represents the expected response when $x_i = 0$; β_i is called the slope and represents the expected increment in the response per unit change in x_i.

In this case, the sklearn.linear regression model developed for Python is used.

The programming code used for linear regression calculates by default the intercept and, obviously, the coefficients of the model, by fitting the model on the base of true training data *x_train* and target value *y_train* arrays, without any normalization. Once the model is fitted, the predicted values array *y_plot* has been obtained on the base of the linear regression model.

In Figure 3.12, the linear regression model of training data, the x_train values versus the x_plot predicted value are shown in the scatter plot for the heating and cooling datasets.

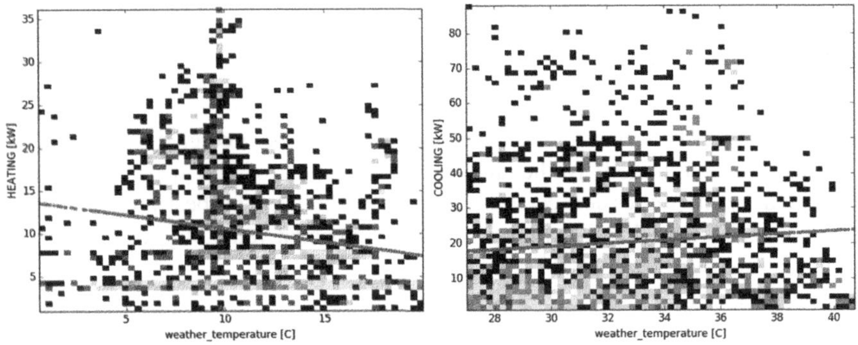

Figure 3.12. *Linear regression model with a scatter plot of training data*

As explained, several metrics are calculated in order to quantify the accuracy of the model. As expected, the scores obtained from metrics in this case are very low, as shown in Table 3.6.

3.2.5.8. *Multiple regression model*

In order to obtain a more accurate model, two different variables are added in the x_train predictor, which in this case becomes a 3D-array with three columns: *external temperature*, *orari* and *deltaT variables* (explained in the previous section). This procedure is called "multiple linear regression analysis" and is used to establish a relationship between the y = y_train array, that is, the energy consumption variable (dependent variable) and the X = x_train(X_1, X_2, X_3) predictor in the form:

$$y = \beta_0 + \beta_1 X_1 + \beta_2 X_2 + \ldots + \beta_n X_n + \varepsilon$$

where X_1 = *external temperature*, X_2 = *orari*, X_3 = *deltaT* and β_0 ... β_n are the regression coefficients to be estimated based on a record of observations and ε is the error. This is normally done by curve fitting based on the least square method with the aim of minimizing the difference between the observed and estimated values. The last term in the equation (ε) is referred to as the residual (or fitted error) and is used for testing the significance of each regression coefficient.

3.2.5.9. *Two variables linear regression*

As previously described, the linear regression model calculates by default the intercept and, obviously, the coefficients of the model, by fitting the model on the base of true training data *x_train*, which in this case is a 2D-array – composed of the *external temperature* and *orari* variables – and target value *y_train* arrays, without any normalization. Once the model is fitted, we can obtain the predicted values array *y_plot* on the base of the linear regression model in the form:

$$y_{pred} = y_{plot} = \beta_0 + \beta_1 X_1 + \beta_2 X_2$$

where X_1 = *external temperature*, X_2 = *orari and* β_0 ... β_2 are the regression coefficients to be estimated based on a record of observations.

In Figure 3.13 the linear regression model is shown. In particular, the external temperature variable versus the x_plot predicted value are represented in the scatter plot of training data, both for the heating and cooling datasets.

Table 3.6. shows the scores obtained from metrics.

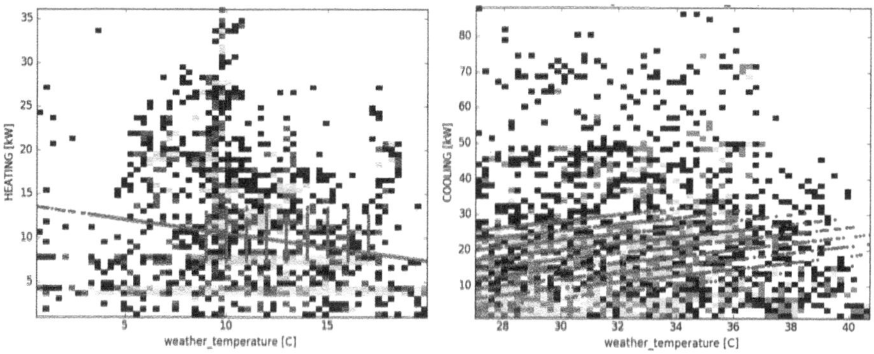

Figure 3.13. *Two variables linear regression model with scatter plot of training data*

3.2.5.10. *Three variables linear regression*

The three variables linear regression model calculates the intercept and the coefficients of the model, by fitting the model on the base of a 3D-array – composed of *external temperature, orari* and deltaT variables – and target value *y_train* arrays, without any normalization. Once the model is fitted, we can obtain the predicted values array *y_plot* on the base of the linear regression model in the form:

$$y\_plot = \beta_0 + \beta_1 X_1 + \beta_2 X_2 + \beta_3 X_3$$

where X_1, X_2, X_3 are external temperature, orari and deltaT and β_0 ... β_3 are the regression coefficients to be estimated based on a record of observations.

In Figure 3.14 the linear regression model is shown. Also, in this case, the external temperature variable versus the *x_plot* predicted value are represented in the scatter plot of training data (*x_train* and *y_train*), both for the heating and cooling datasets.

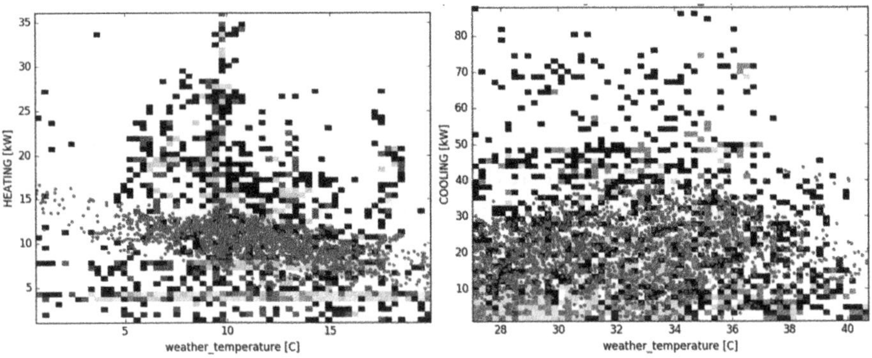

Figure 3.14. *Three variables linear regression model with scatter plot of training data*

Table 3.6. shows the scores obtained from metrics. By fitting the model with the additional variables, each metric gives slightly better results, but the model, as expected, is still far from accurate. For these reasons, a different regression technique will be used and explained in the next section.

	HEATING REGRESSION MODEL			COOLING REGRESSION MODEL		
Parameters	1_variable – external temperature	2_variables – external temperature and orari	3_variables – external temperature, orari, deltaT	1_variable – external temperature	2_variables – external temperature and orari	3_variables – external temperature, orari, deltaT
Coefficients β3	-	-	-0.70321422	-	-	3.255
Coefficients β2	-	-0.397	-0.44475693	-	-2.711	-2.459
Coefficients β1	-0.320	-0.352	0.30043262	0.434	0.928	1.628
Intercept β0	13.744	19.100	7.59815274	6.179	24.754	81.682
Explained variance score	0.03	0.05	0.07	0.01	0.13	0.29
Mean squared error	48.13	47.27	45.98	270.91	236.27	193.18

Root mean squared error	5.66	5.57	5.34	12.77	11.43	10.45
Median absolute error	5.35	5.12	4.72	11.43	8.54	8.28
R2 score	0.03	0.05	0.07	0.01	0.13	0.29
MAPE %	86.68%	85.42%	80.52%	177.16%	155.67%	125.99%

Table 3.6. *Metrics results*

3.2.5.11. *Ridge regression*

Ridge regression (also known as Tikhonov regularization or, in machine learning, as weight decay) is related to the Levenberg–Marquardt algorithm for non-linear least-squares problems. This method also has built-in support for multivariate regression, and it solves a regression model where the loss function is the linear least squares function and regularization is given by the l2-norm regularization.

The equation with A matrix and b vector

$$Ax = b$$

is usually solved with the ordinary least squares linear regression. Ordinary least squares seek to minimize the sum of squared residuals, which can be written as:

$$||Ax = b||^2$$

The ridge regression model is solved by including a regularization term in this minimization:

$$||Ax = b||^2 + ||\Gamma x||^2$$

where Γ is the Tikhonov matrix (Naiyang 2012). In many cases, this matrix is chosen as a multiple of the identity matrix giving preference to solutions with smaller norms; this is known as $L2$ regularization (Pedregosa *et al.* 2011). The $L2$ loss function minimizes the sum of the square of the differences between the true value and the estimated values as:

$$L2 = \sum_{i=1}^{n} \left(y_{i\,train} - y_{i\,pred}\right)^2$$

where $y_{i\,train}$ is the i true value and $y_{i\,pred}$ the estimated i values.

As explained in Scikit-learn (Pedregosa *et al.* 2011), the ridge regression procedure addresses some of the problems of ordinary least squares by imposing a penalty on the size of coefficients. The ridge coefficients minimize a penalized residual sum of squares introducing α, a fixed positive constant that has to be tuned in order to obtain a good calibration of the penalty term. α is a complexity parameter that controls the amount of shrinkage and therefore regularization strength. Regularization improves the conditioning of the problem and reduces the variance of the estimates. Larger values specify stronger regularization.

When α = 0 a classical linear regression is obtained.

In order to validate α, a built-in cross-validation is used. The *RidgeCV scklearn* command implements ridge regression with built-in cross-validation of the alpha parameter. Then α is obtained by creating a function that tests each value of α by fitting *x_train, y_train* and checking the score on the base of the mean absolute error. The α value obtained is equal to 0.005.

3.2.5.12. *Single variable linear regression*

In this case, the *sklearn* polynomial features and ridge regression estimator developed for Python are used in order to consider a second-degree model. Since a second-degree model is used, the regression model to express the dependence of the expected response on the predictor can be written as:

$$Y_{pred} = \beta_0 + \beta_1 x^2 + \beta_2 x$$

where $Y_{pred} == y_{plot}$ is the predicted value and $x = x_{train}$ is the *external temperature* variable.

As previously described, α is equal to 0.005 and an automatic solver is used based on the type of data. This automatic solver can choose between the "*singular value decomposition* of X" to compute the ridge *coefficients*, the "*standard scipy.linalg.solve*" function to obtain a closed-form solution, the "*sparse_cg*", which uses the conjugate gradient solver as found in *scipy.sparse.linalg.cg*, or the regularized least-squares routine, "*sag*", which uses a stochastic average gradient descent.

Also, in this case, it is possible to extract the intercept and obviously, the coefficients of the second-degree model, by fitting the model on the base of true training data *x_train* and target value *y_train* arrays, without any normalization. Once the model is fitted, the predicted values array *y_plot* can be obtained on the base of ridge regression model.

In Figure 3.15, the non-linear regression model that is the *x_train* values versus the *x_plot* predicted value in the scatter plot of training data is shown (*x_train* and *y_train*) both for the heating and cooling datasets.

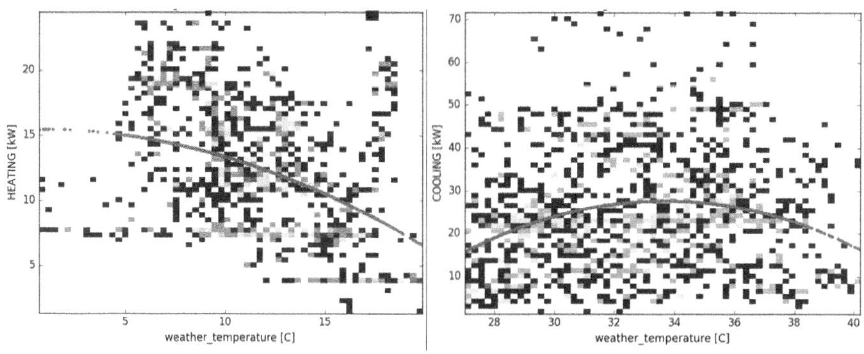

Figure 3.15. *Ridge second order degree regression model with scatter plot of training data*

In this case, the same metrics previously described are calculated in order to quantify the accuracy of the model (Table 3.7).

3.2.5.13. *Two variables linear regression*

By adding the *orari* variable, *x_train* becomes a 2D-array composed of the external temperature array and *orari* array. The regression model to express the dependence of the expected response on the predictor can be written as:

$$Y_{pred} = \beta_0 + \beta_1 x^2 + \beta_2 y^2 + \beta_3 xy + \beta_4 x + \beta_5 y$$

where $Y_{pred} = y_{plot}$ is the predicted value, predictor x is the external temperature, predictor y is the *orari* variable, and β_0 ... β_5 are the regression coefficients to be estimated based on a record of observations.

Figure 3.16 shows the scatter plot of a non-linear regression model obtained by fitting two variables, both for the heating and cooling datasets, by considering in x-axis the ambient temperature.

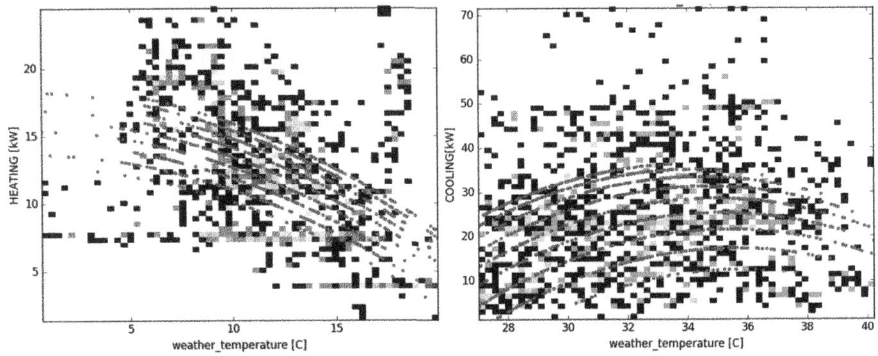

Figure 3.16. *Ridge second order degree regression model with a scatter plot of training data*

The same metrics are being calculated in order to quantify the accuracy of the model (Table 3.7).

3.2.5.14. *Three variables linear regression*

Finally, also in this case, by adding the *deltaT* variable, *x_train* becomes a 3D-array composed of *external temperature array*, *orari* and *deltaT* arrays. The regression model to express the dependence of the expected response on the predictor can be written as:

$$Y_{pred} = \beta_0 + \beta_1 x^2 + \beta_2 y^2 + \beta_3 z^2 + \beta_4 xy + \beta_5 yz + \beta_6 zx + \beta_7 x + \beta_8 y + \beta_9 z$$

where $Y_{pred} = y_{plot}$ is the predicted value, predictor x is the external temperature, predictor y is the orari variable, z deltaT variable and β_0 ... β_9 are the regression coefficients to be estimated based on a record of observations.

Figure 3.17 shows the scatter plot of the non-linear regression model obtained by fitting three variables, both for heating and cooling datasets, by considering in x-axis the weather temperature.

The same metrics are calculated in order to quantify the accuracy of the model as shown in Table 3.7. By considering a non-linear regression model, the accuracy of the results drastically increases.

Once the coefficients are obtained, the equations related to load consumptions, both for heating and cooling, are composed and integrated in the TRNSYS model, in order to obtain an initial estimation of the NTLab loads. As previously explained, this procedure was chosen in a preliminary stage, to avoid spending a lot of time carrying out a simulation model of a smart grid plan where *Out_temperature* is the external temperature channel, *Time_hour* is the hour of the day, and *Room* is *deltaT*, the difference between external and internal temperature, explained in the previous paragraph.

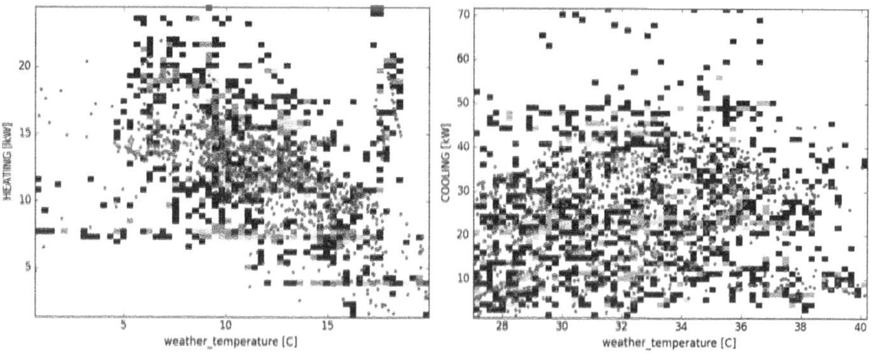

Figure 3.17. *Second order ridge regression model with scatter plot of training data*

In conclusion, there is still space for improvement, considering the accuracy of the regression model, since the MAPE metric is around 27% for both datasets, but the time efficiency of this approach once the code is compiled in Python presents a significant advantage. So, if the goal is a preliminary estimation of building consumptions, this approach is very powerful because one can check the results very fast and obtain an estimate of the loads required by the building in varying conditions.

The energy signature obtained with a multiple ridge regression model can be a good compromise in a preliminary stage, when one needs to have a general overview of the smart grid nodes that will compose the smart grid and to understand what the loads are as well as the production that is necessary for energy balance.

Further work is necessary in the areas of data collection to enable benchmarks to be defined, and practical applications are needed in order to improve reliability or reveal limitations. For example, by using occupancy data and distinguishing two types of days (occupied or unoccupied), the accuracy of the model can be greatly improved (Rabl and Rialhe 1992).

	HEATING REGRESSION MODEL			**COOLING REGRESSION MODEL**		
	1_variable – external temperature	2_variables – external temperature and orari	3_variables – external temperature, orari, deltaT	1_variable – external temperature	2_variables – external temperature and orari	3_variables – external temperature, orari, deltaT
Coefficients β9	-	-	0.027	-	-	-0.476
Coefficients β8	-	-	-0.197	-	-	-0.569
Coefficients β7	-	-	0.016	-	-	-0.037
Coefficients β6	-	-	-0.435	-	-	1.059
Coefficients β5	-	-0.027	0.178	-	-0.271	0.612
Coefficients β4	-	0.004	0.367	-	0.114	-0.757
Coefficients β3	-	-0.026	7.386	-	-0.248	-17.680
Coefficients β2	0.024	-0.067	-4.474	-0.267	0.065	-17.778
Coefficients β1	0.032	-0.065	-13.885	18.007	15.832	32.975
Intercept β0	15.432	21.610	149.746	-275.658	-228.454	-308.487
Root mean squared error	3.750	3.49	**2.840**	10.350	8.280	**6.270**
R2 score	0.160	0.27	**0.520**	0.060	0.340	**0.640**
MAPE	39.72%	35.60%	**27.28%**	73.65%	53.69%	**27.43%**

Table 3.7. *Metrics results*

3.3. Linear Fresnel solar collector at the NTL building, Cyprus Institute[2]

The National Technologies Laboratory (NTL) (Figure 3.18) of Cyprus Institute is a top-end tertiary building, which incorporates energy efficient technologies and integrates modern HVAC (heating, ventilation, air conditioning) technology with a rooftop linear Fresnel solar collector (Figure 3.19), significantly reducing energy demand from the power grid. The installation of the new solar cooling plant at the NTL premises has been supported by the STS-Med project, financed by the ENPI CBCMED program of the European Union, where it has been part of a set of four pilot plants which should demonstrate the potential application of solar energy in a built environment, with a deep integration with the demand of energy raised by the neighborhood (Montenon *et al.* 2017).

Figure 3.18. *NTL building*

The building was completed in 2014 with the goal of minimizing consumption for heating and cooling, and producing part of its electrical needs from a building integrated PV system. The baseline scenario concerned heating and cooling loads of

2 This section is based on the following article, used with permission from Elsevier: Nikos KAMPELIS, Konstantinos GOBAKIS, Vagias VAGIAS, Denia KOLOKOTSA, Laura STANDARDI, Daniela ISIDORI, Cristina CRISTALLI, Fabio Maria MONTAGNINO, Filippo PAREDES, Pietro MURATORE, Luca VENEZIA, Marina Kyprianou DRACOU, Alaric MONTENON, Andri PYRGOU, Theoni KARLESSI and Mat SANTAMOURIS (2017). "Evaluation of the performance gap in industrial, residential and tertiary near-zero energy buildings", *Energy and Buildings*, 148, 58–73.

61.2 kWh/m^2 and 78.3 kWh/m^2 respectively. The design of interventions was based on simulations used to test various effects such as heat losses, solar gains, ventilation (free/night cooling, demand control) and daylight availability, taking into account external shading constructions. The various interventions proposed during the initial design phase were double glazing, a high thermal resistance building envelope, a green roof (50% coverage), shading, ceiling fans, high efficiency, waste heat recovery, ventilation and ground heat exchangers. At the design stage it was considered that the above measures would result in 83% and 61% reduction of heating and cooling loads, which translates to 10.7 kWh/m^2 and 30.2 kWh/m^2 respectively. The above figures were obtained on the basis of international standards (EN 15251) for indoor comfort, including the fresh air ventilation rate, indoor temperature, indoor relative humidity and illuminance levels. The equivalent primary energy consumption for heating and cooling is 28.9 kWh/m^2 and 81.54 kWh/m^2 respectively (Fokaides *et al.* 2011). Moreover, it was estimated that the energy balance could further be improved by the installation of a PV in an area of 120m^2 to produce 21MWh (9.58 kWh/m^2) on a yearly basis.

Figure 3.19. *Linear Fresnel solar collector*

The Novel Technologies Laboratory is an educational/research building mainly containing offices and laboratories, which operates 8 hours/day, Monday to Friday and for about 240 days/year. Cooling and heating of the building are provided by three air-cooled chillers with partial heat recovery connected to fan coil units (FCUs) in office areas and air handling units (AHUs) in laboratories and the lecture hall. Two of them are cooling with just a capacity of 243.27 kW with EER = 2.35, ESEER = 4.17, while the other one has a cooling capacity of 234.29 kW with EER = 2.24 and ESEER = 3.97 and a heating capacity of 293 kW with COP = 3.15.

Ventilation at offices is achieved with local heat recovery units (HRUs), at the lecture hall and laboratories with AHUs and the WC areas with local extract fans. The six AHUs are supplied with hot and cold water, generated by the two chillers and the heat pumps. Each unit has a heat recovery unit where indoor air and fresh air are mixed. Fresh air flow is adjusted by motorized dampers which are controlled by temperature and indoor air quality sensors. Moreover, the AHUs have a free-cooling function to increase the supply of fresh air if the outdoor air temperature is lower than the indoor temperature. AHUs are manually controlled by local controllers in spaces, and they are connected to the BEMS (building energy management system) from where they are monitored and controlled as well. AHUs have a nominal electric power of 44.300 W and a nominal air flow of 49.310 m^3/h.

Heat recovery units (HRUs) extract air from the return grilles of rooms and supply fresh air at the return of FCUs to be cooled or heated. HRUs are controlled by local air-conditioning controllers in offices, and they are connected to the BEMS from where they are monitored and controlled as well. HRUs have a nominal electric power of 1.972 W and a nominal air flow of 4.960 m^3/h.

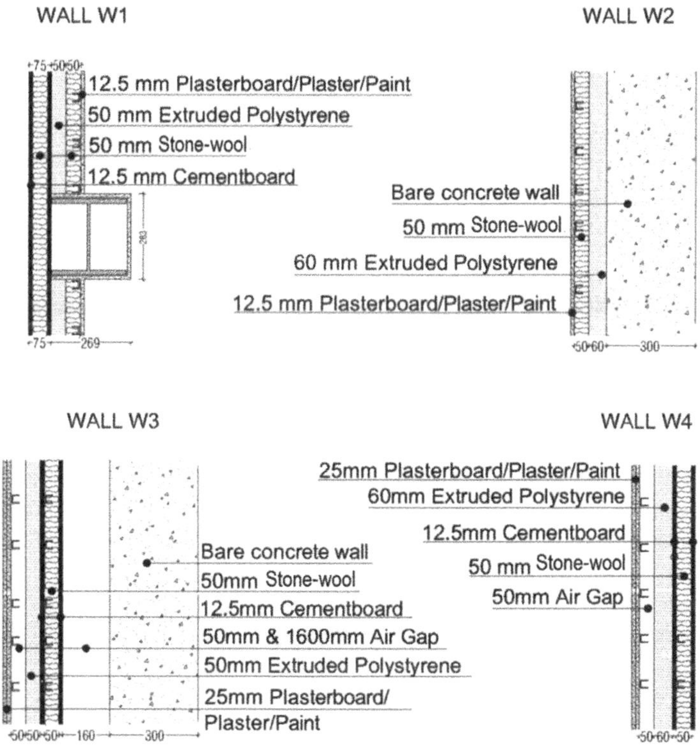

Figure 3.20. *Different wall structures and materials applied in the NTL building*

Lighting is achieved with compact fluorescent lamps, T8 fluorescents and metal halide lamps, which are controlled by local on/off switches. Total lightning power is 41.700 W.

Four different wall types W1–W4 have been implemented as shown in Figure 3.20 with U-valueWALL1 = 0.249 W/m2K, U-valueWALL2 = 0.305 W/m2K, U-valueWALL3 = 0.294 W/m2K, U-valueWALL4 = 0.295 W/m2K respectively (Fanos 2014).

Both extruded polystyrene and stone-wool have been used for thermal insulation. Also, for the windows and glass doors, double glazing has been used. The glazing consists of 4 mm of low emittance glass – 16 mm air – 4 mm low emittance glass with Ug-value = 1.4W/(m^2K), light transmission 71% and solar factor 42%. Moreover, at the east and west side of the building, perforated galvanized sheet iron has been installed with 60% percentage of perforation.

A rooftop installation of 62 panels, 180 Wp each (11.16 kWp total) with a small installation of 54 biPV shadowing panels, 62 Wp each (3.348 kWp total) in the south façade partially covers the electricity consumption of the NTL.

During energy auditing, it has been observed that the initial design of the building was aimed to ensure minimum energy consumption: insulation and shadowing have been introduced, efficient lamps have been used for lighting and controls for ventilation, and heat recovery units and air flow have been installed.

However, geothermal and solar thermal systems have not been envisaged from the beginning, neither have lighting control systems and intelligent control of the air flow in each room or section.

The introduction of an innovative system has been taken into account and, in particular, so has the installation of an integrated solar cooling system in an advanced state under the above-mentioned STS-Med project. This pilot plan at the Cyprus Institute relies on a Fresnel collector that has been installed on the rooftop of a school just in front of the NTL building.

This collector has been specifically designed by Idea Srl to assure the independent rotation of each collecting element of the primary optics. This allows the complete rotation of the supports that can host in future a PV panel on the back to provide electrical energy in the absence of a suitable direct radiation for thermal energy exploitation.

The technical specifications and main features of the linear Fresnel collector connected to the Novel Technologies Laboratory are listed in Table 3.8.

Element	Characteristics
Mirrors	288 mirrors (2,000 mm x 320 mm dimension per unit) Global aperture area of 184.32 m^2 Three focusing distances according to the position of each row 144 collecting units distributed into18 rows x 8 modules
Motorization	72 DC motors (1 for a couple of collecting units) controlled by 18 PLC/36 drivers
Receiver	32 m long Secondary reflector on top Vacuum tube receiver
Weight	8 tons distributed on 9 beams
Orientation	Receiver is aligned with the local meridian
Heat transfer fluid	Diathermic oil (Duratherm 450)
Maximum thermal power @850W/m² DNI	70 kW 170°C outlet nominal temperature

Table 3.8. *Main features of the installed LFR system*

The main thermal loops that have been implemented are:

1) Thermal oil loop: the heating medium which circulates inside the Fresnel collector, equipped with a small buffer to preheat the oil and Nitrogen pressurized (3 bar) system in order to maintain the oil above a minimum pressure.

2) Thermal storage loop: pressurized water, able to store the heat produced by the Fresnel collector for up to 2 hours for cooling and 4 hours for heating; an expansion tank maintains the water at high pressure (up to 8 bar).

3) Heating-medium loop: hot water to fire the absorption chiller, equipped with a small buffer.

4) Heat recovery loop: hot water to support heating in winter.

5) HVAC chiller loop: a chilled circuit that is connected to the previously implemented HVAC system of the Novel Technologies Laboratory.

A schematic representation of the loop is provided in Figure 3.21.

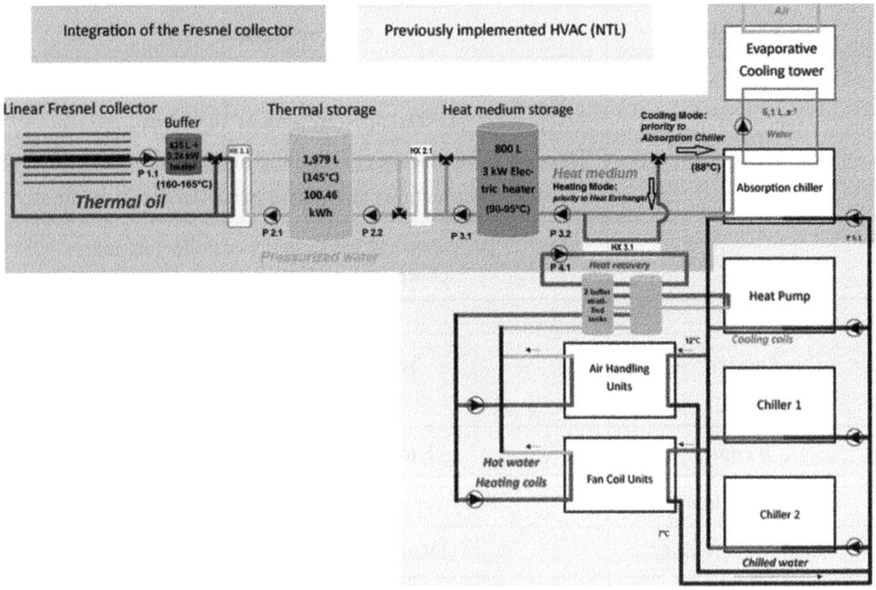

Figure 3.21. *Layout of the SHC (solar heating and cooling) system and integration with the HVAC system in the NTL*

3.3.1. *Development of the NTL model*

The thermal energy annually provided by the solar collector has been evaluated, using an hourly steady-state model that simulated the annual solar energy gain.

A tool has been developed by the authors (using a spreadsheet) to calculate the hourly thermal energy produced by the LFR collector. Starting from hourly values of an astronomical coordinate relative to the Sun's position (azimuth, elevation) generated using the SOLRAD (Earth System Research Lab 2021) v 16.0 tool, the optical features of the LFR, like the incident angle modifier (IAM transversal, IAM longitudinal), the "end effect" factor and the optical efficiency, were calculated.

The weather data, used for the hourly simulation, were generated by Meteonorm version 7.1 (Meteonorm n/a) with an interpolation mode from the hourly base.

Thermal losses of absorber pipes were calculated, according to equations applicable to the concentrating solar collector, as defined by Kalogirou (2004) and UNI EN ISO 9806 (Solar Thermal Collectors: Test Methods). Results in terms of thermal energy produced by LFR, calculated through a steady-state model instead of a quasi-dynamic model, could be affected by an overestimation error lower than 10% (Nicole Janotte 2009). Thus, values of thermal energy gain have been reduced by 10%, to correct this error.

To estimate the heating and cooling energy that could be supplied by the single-effect absorption chiller system, the thermal capacity of the storage system in operative conditions was calculated. According to this analysis, as it was not possible to use an energy building model to describe the hourly thermal needs of the building, values of electric energy consumption provided by the metering system installed in the building were considered. Every 15 minutes, the metering system collects the cumulative consumption data for lighting, power and HVAC for each floor, as well as the generation of electricity by the PV system installed on the rooftop. Full datasets available for the years 2014 and 2015 were exploited.

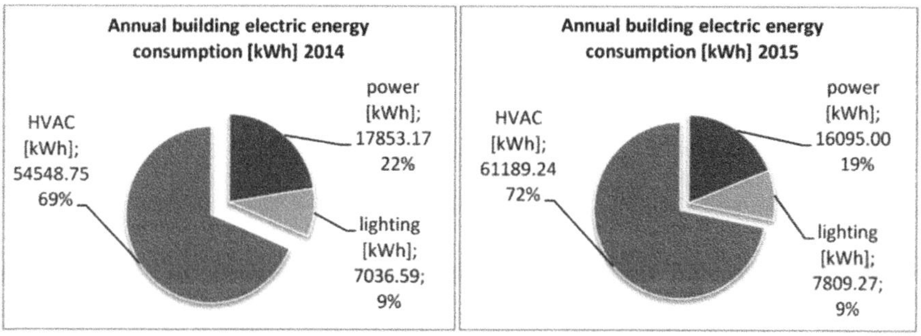

Figure 3.22. *Categorized electrical energy consumption [kWh, %] of the NTL in 2014 and 2015*

The two diagrams of Figure 3.22 show how a large part of the electric energy consumption is attributable to the HVAC system, respectively 69% in 2014 and 72% in 2015, with an increase of 6,500 kWh in one year. We note that the energy for specialised research-related activities of the laboratories is not accounted for in these diagrams. The installed PV system has covered the 27% of the annual electric energy consumption of the building (offices, without laboratories). This percentage is almost the same in both years. The increase is partially due to the increased number of employees from 2014 to 2015, as well as because of the heavier climatic conditions in 2015.

The electrical energy consumptions in 2014 and 2015, due to the HVAC system, were assigned to Heating and Cooling and normalized, considering the HDD (heating degree days) (base temperature = 19°C) and CDD (cooling degree days) (base temperature = 26°C), calculated using data collected by the closest available weather station (LCNC – 17607 – Athalassa, CY – 33.40E; 35.14N) (Table 3.9). The base temperatures were chosen in accordance with the study conducted by Theophiloua and Serghides (2015).

	Electric energy consumption in heating (HVAC) [kWhe]	HDD [°C*Day] (base temperature = 19°C)	Normalized electric energy consumption [kWhe/ HDD]	Electric energy consumption in cooling (HVAC) [kWhe]	CDD [°C*Day] (base temperature = 26°C)	Normalized electric energy consumption [kWhe/ CDD]
2014	22783.04	692.7	32.9	31765.71	684.8	46.4
2015	24774.91	788.1	31.4	36224.47	810	44.7
Variation [%]	9%	14%	-4%	14%	18%	-4%

Table 3.9. *NTL electric energy consumption in 2014 and 2015 due to a HVAC system, normalized with respective HDD (heating degree days) and CDD (cooling degree days). For a color version of this table, see www.iste.co.uk/kolokotsa/smartbuildings2.zip*

3.3.2. *Energy performance analysis in the NTL*

The impact of the solar cooling system has been simulated in two scenarios. The first one assumes the nominal optical efficiency of the LFR solar field at 46%. In the second scenario, the optical efficiency was set to 60% achievable with a potential improvement of the control of the mirrors.

Figure 3.23 shows the thermal energy gained monthly by the LFR system in the first scenario (optical efficiency at 46%). Table 3.10 shows the main values of thermal energy production.

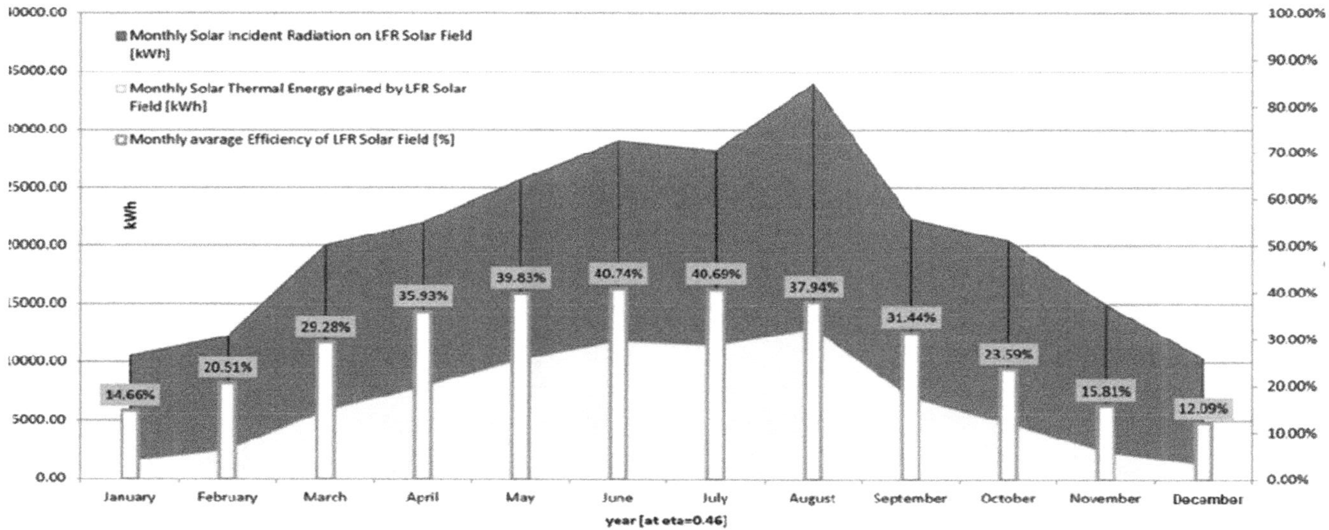

Figure 3.23. *Histogram of monthly thermal energy gained by the LFR system and monthly average efficiency in Scenario 1 (optical efficiency = 46%)*

Month	**Monthly solar thermal energy gained by LFR solar field [kWh]**	**Monthly solar incident radiation on LFR solar field [kWh]**	**Monthly average efficiency of LFR solar field [%]**	**Peak power [KW]**	**Monthly thermal peak efficiency of LFR [%]**
January	1,548.8	10,567.7	14.6%	25.7	19.2%
February	2,509.7	12,235.2	20.5%	35.9	27.2%
March	5,858.9	20,006.9	29.3%	53.0	35.3%
April	7,895.9	21,977.7	35.9%	69.1	40.4%
May	1,225.4	25,674.5	39.8%	64.5	43.2%
June	11,807.1	28,983.0	40.7%	70.1	43.8%
July	11,459.41	28,164.9	40.7%	71.5	43.8%
August	12,903.6	34,013.7	37.9%	62.8	42.1%
September	7,010.2	22,294.9	31.4%	51.3	37.4%
October	4,819.3	20,431.5	23.6%	45.4	30.2%
November	2,363.6	14,951.6	15.8%	28.6	21.2%
December	1,247.4	10,320.1	12.1%	19.5	14.6%
Total year	**79,649.4**	**249,622.3**	**31.9%**	**Optical efficiency = 46%**	

Table 3.10. *Main monthly thermal energy features of LFR production in Scenario 1. For a color version of this table, see www.iste.co.uk/kolokotsa/smartbuildings2.zip*

In the second scenario, presented in Figure 3.24 and Table 3.11 (optical efficiency at 60%), the annual solar thermal energy production could be increased by more than 30%, reaching approximately 104 MWh.

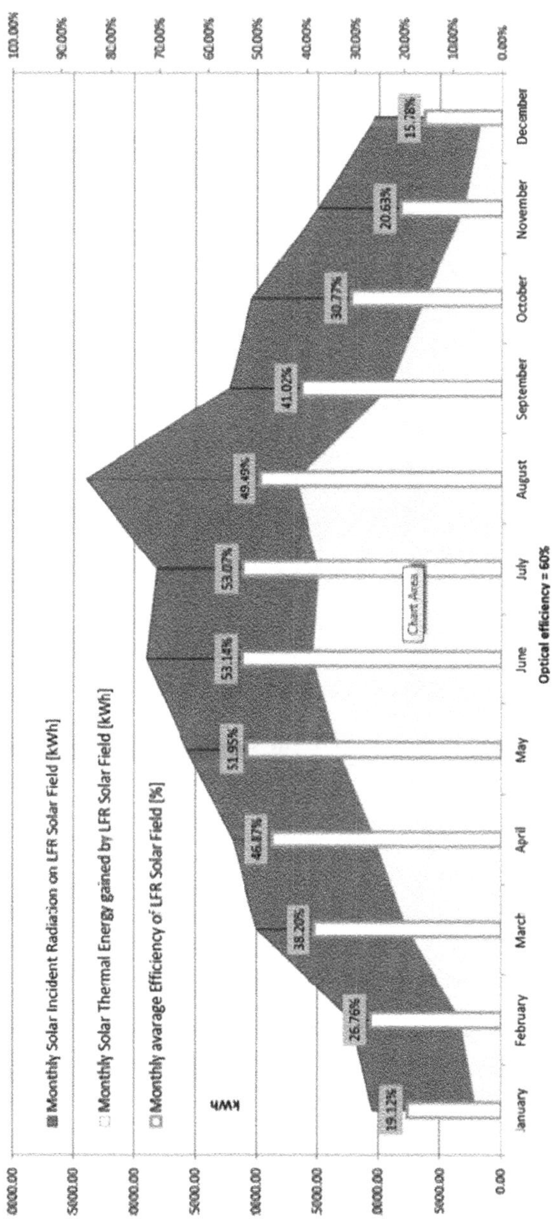

Figure 3.24. *Histogram of monthly thermal energy gained by the LFR system and monthly average efficiency in Scenario 2 (optical efficiency = 60%)*

Month	Monthly solar thermal energy gained by LFR solar field [kWh]	Monthly solar incident radiation on LFR solar field [kWh]	Monthly average efficiency of LFR solar field [%]	Thermal peak power [KW]	Monthly thermal peak efficiency of LFR [%]
January	2,020.9	10,567.7	19.1%	33.6	25.0%
February	3,274.4	12,235.2	26.7%	46.8	35.5%
March	7,643.2	20,006.9	38.2%	69.2	45.9%
April	10,300.0	21,977.8	46.8%	90.2	52.7%
May	13,338.7	25,674.5	51.9%	84.2	56.3%
June	15,401.6	28,983.0	53.1%	91.5	57.2%
July	14,948.1	28,164.9	53.1%	93.2	57.1%
August	16,832.1	34,013.8	49.5%	81.9	54.8%
September	9,144.8	22,294.9	41.0%	66.9	48.8%
October	6,286.9	20,431.6	30.8%	59.3	39.4%
November	3,083.9	14,951.7	20.6%	37.4	27.6%
December	1,628.1	10,320.2	15.8%	25.5	19.1%
Total Year	***103,902.87***	***249,622.3***	***41.6%***	***Optical efficiency = 60%***	

Table 3.11. *Main monthly thermal energy features of LFR production in Scenario 2. For a color version of this table, see www.iste.co.uk/kolokotsa/smartbuildings2.zip*

To compare the share of the SHC in covering the thermal energy building demand in each scenario (Table 3.12), the following assumption were made:

– 10% of thermal energy losses in piping systems, storage systems;

– thermal COP = 1.0 in heating mode, and COP = 0.7 in cooling mode for the SHC system;

– EER = 2.3 and SEER = 4 in cooling mode, and COP = 3.15 in heating mode for chillers and heat pumps of the HVAC system;

– the hourly electric consumption of the SHC is set to 3.5 kWh in cooling mode and 2.0 kWh per hour in heating mode (since in heating mode, the absorption chiller and the cooling tower are OFF).

The actual COP of each system could not be calculated because the external air temperature and the hourly thermal energy needs of the building have not been registered. As for the scope of this work, only the nominal values were considered. For the purpose of further investigation and more accurate simulations, temperature sensors of outdoor temperature should be installed.

2015	Month	Monthly SHC capacity [kWh]	Monthly HVAC thermal supply [kWh]	Building thermal needs covered by SHC [%]	Monthly HVAC electric energy consumptions [kWh]	Monthly HVAC electric energy consumptions with SHC [kWh]	Monthly SHC electric energy saving [kWh]	HVAC electric energy saving due to SHC [%]
Scenario 1 (Optical efficiency at 46%)	*January*	1,471.3	19,819.8	7.4%	6,384.8	5,949.6	435.1	6.8%
	February	2,384.2	18,785.4	12.7%	5,942.2	4,997.5	944.6	15.9%
	March	5,259.9	11,012.1	47.8%	3,374.9	2,134.1	1,240.7	36.7%
	April	6,172.6	7,743.9	79.7%	2,237.2	821.7	1,415.5	63.2%
	May	6,896.0	11,060.2	62.4%	3,151.5	1,163.0	1,988.4	63.1%
	June	7,851.7	19,320.8	40.6%	5,292.4	2,476.6	2,815.7	53.2%
	July	7,620.5	34,985.8	21.8%	8,315.0	5,966.7	2,348.2	28.2%
	August	8,580.9	34,022.3	25.2%	9,259.4	5,571.1	3,688.3	39.8%
	September	4,661.7	34,200.6	13.6%	7,968.7	6,715.0	1,253.7	15.7%
	October	3,233.4	12,927.	25.0%	3,345.3	2,782.3	563.0	16.8%
	November	1,874.7	5,446.6	34.4%	1,784.1	1,550.4	233.7	13.1%
	December	1,135.2	13,264.9	8.6%	3,943.4	3,803.1	140.2	3.5%
	Annual	***57,142.4***	***222,589.8***	***25.7%***	***60,999.3***	***43,931.7***	***17,067.5***	***27.9%***
Scenario 2 (Optical efficiency at 60%)	*January*	1,919.8	19,819.8	9.7%	6,384.8	5,743.2	641.61	10.0%
	February	3,110.6	18,785.4	16.6%	5,942.2	4,631.7	1,310.5	22.0%
	March	6,861.8	11,012.1	62.3%	3,374.9	1,634.6	1,740.2	51.5%
	April	8,052.0	7,743.9	104.0%	2,237.2	210.6	2,026.5	90.5%
	May	8,995.7	11,060.2	81.3%	3,151.5	305.2	2,846.2	90.3%
	June	10,242.1	19,320.8	53.0%	5,292.4	1,353.1	3,939.3	74.4%
	July	9,940.4	34,985.8	28.4%	8,315.0	4,991.8	3,323.1	39.9%
	August	11,193.3	34,022.3	32.9%	9,259.4	4,144.5	5,114.9	55.2%
	September	6,081.2	34,200.6	17.8%	7,968.7	6,115.8	1,852.9	23.2%
	October	4,218.1	12,927.0	32.6%	3,345.5	2,415.9	929.4	27.7%
	November	2,446.1	5,446.6	44.9%	1,784.1	1,354.8	429.3	24.0%
	December	1,481.7	13,264.9	11.2%	3,943.4	3,687.5	255.8	6.4%
	Annual	***74,543.2***	***222,589.8***	***33.5%***	***60,999.3***	***36,589.2***	***24,410.1***	***40.0%***

Table 3.12. *Main monthly results of the simulations in reference to 2015, in terms of cooling and heating energy provided by SHC and electric energy saving effect on the building for each scenario*

In Scenario 1, the SHC system could yearly cover about of 26% of thermal needs of the building, with an average energy saving effect on the HVAC electric energy consumptions of about 28% (Figure 3.25).

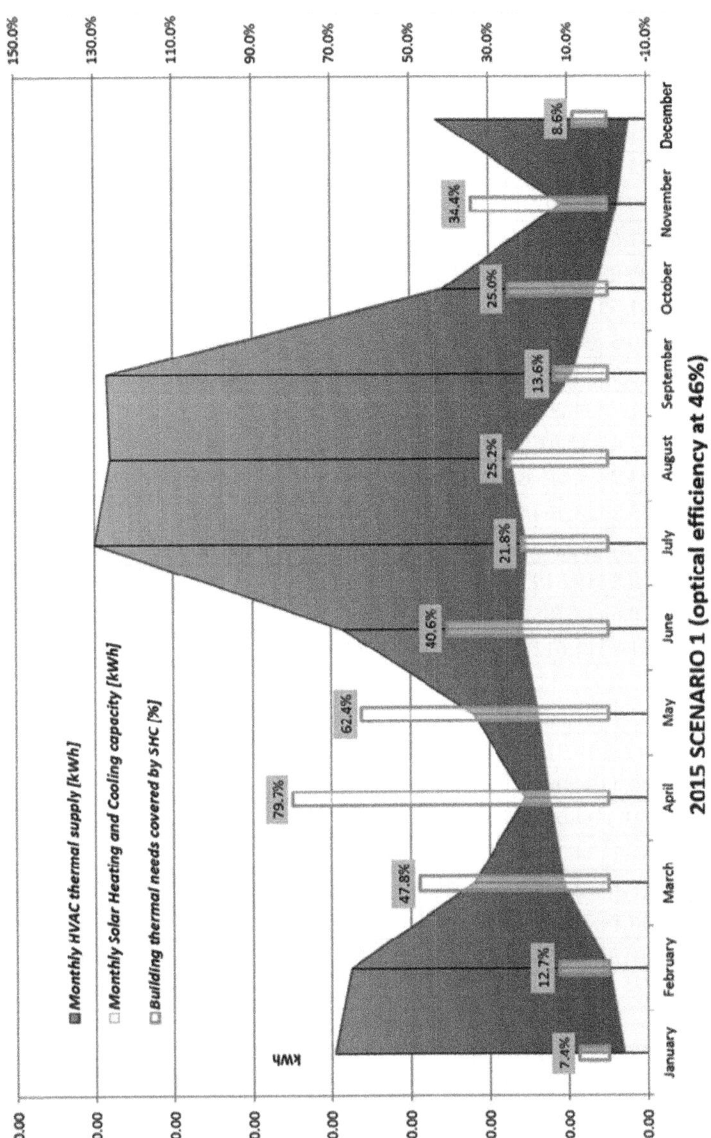

Figure 3.25. *NTL monthly thermal demand covered by the SHC system in Scenario 1*

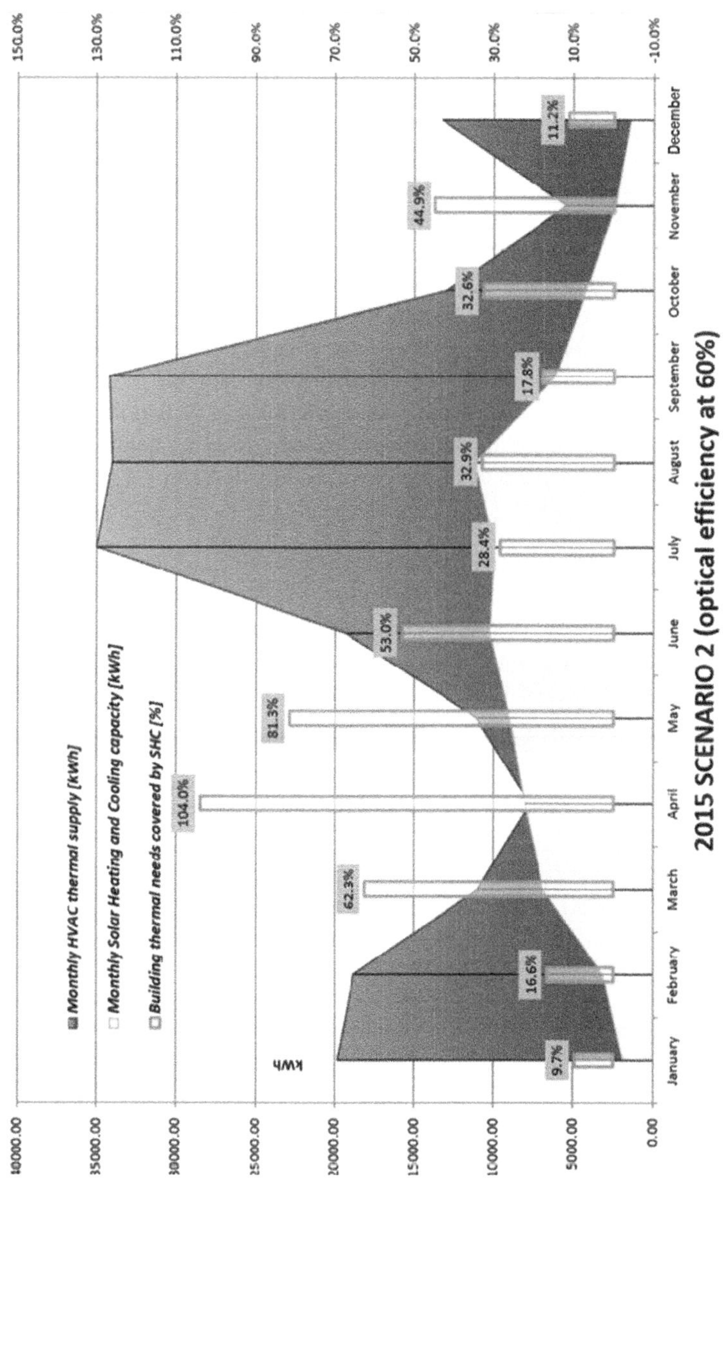

Figure 3.26. *NTL monthly thermal demand covered by the SHC system in Scenario 2*

In Scenario 2, the SHC rate could be higher than 33% regarding thermal needs of the building with an energy saving effect on HVAC electric consumptions of about 40% (Figure 3.26). In this case, the simulations show that the absorption chiller capacity should be improved by 30% to get the full benefit of the energy collected by the solar field. During the spring, the SHC is able to cover about 70% of the thermal energy needs of the NTL building. According to Scenario 2, this value reaches 100% in the month of April.

An important improvement of the SHC system has been envisaged by (1) introducing a double-effect absorption chiller, instead of a single-effect one, and (2) covering the back side of LFR mirrors structures with PV panels able to convert diffuse solar radiation, when the direct one is too poor for concentration and when the building doesn't need cooling or heating services. In this configuration, the LFR could work in thermal mode during working days with a DNI higher than 300 W/m^2 and it could be switched in PV mode if the DNI is lower than 300 W/m^2, as well as during weekend or holidays. According to preliminary calculations, this smart configuration could increase the annual amount of energy saving contribution of the LFR solar field from 28% to 38% of the electrical energy demand.

3.3.3. *Discussion*

In the selected case study, the performance gap is finally assessed by comparing the design and operational primary energy consumption in design and operational primary energy consumption in Table 3.13.

Pilot case study	Normalized primary energy consumption in design phase (kWh/m^2)	Normalized net primary energy consumption in operational phase (kWh/m^2)
NTL	64.4	62.9

Table 3.13. *Normalized primary energy consumption in the design and operational phase*

With regards to the NTL building, taking into account initial design assumptions results in an insignificant performance gap. This is especially true when energy consumption based on HVAC and lights is considered excluding power consumption of laboratories. It is to be evidenced that the most relevant effect of the SHC introduction is an improved seasonal balance between demand and generation from renewable sources, smoothing the summer peak of electrical demand for air conditioning.

In the case of the NTL, renovation design and implementation activities were consistent and effective in setting energy efficiency targets, reshaping the envelope and integrating smart controls as well as the linear Fresnel reflector system into the building service operations.

3.4. Conclusion

In this chapter, the operational performance of a building used for research and educational purposes has been investigated, analyzed and optimized with the use of quasi-dynamic simulation. Energy efficient technologies, renewable energy technologies and storage have been audited to evaluate their significance. Various performance indicators have been used in this analysis including normalized primary energy consumption at the design and operational stage. The above analysis reveals the significance in evaluating the actual performance gap in near zero-energy buildings and provides the basis for decision-making and smart adjustments as necessary. Apart from the high-quality building envelope, the near-zero target is largely pursued by renewable energy technologies and the implementation of advanced monitoring and controls. Furthermore, there needs to be a systematic and continuous approach in establishing near zero-energy targets through research and innovation activities.

4

On the Comparison of Occupancy in Relation to Energy Consumption and Indoor Environmental Quality: A Case Study

4.1. Introduction

The energy performance of all dwellings and the quality of the indoor environment has been the topic of research of many groups all over the world. This is of importance for many countries with tropical climatic conditions, where buildings account for over 36% of the total electricity sales in recent years and this percentage is expected to grow in the future (Kwong *et al.* 2014). The present research focuses on the case study of a typical building which operates on the campus of the National University of Singapore (henceforth NUS). Singapore Island experiences a warm and moist climate with a rate of rainfall that averages 2,381 mm per annum. The diurnal variation of temperature is low and ranges by about 7°C. At the same time the mean annual relative humidity is about 84%, with typical daily maxima approaching saturation in the cooler early mornings (de Dear *et al.* 1991). This type of climate may have an adverse impact on occupant comfort (Wong and Khoo 2003). Similarly, the occupants' number, alongside their environmental adaptation and activity in the building, their comfort and the energy consumption, is

---

Chapter written by Margarita Niki ASSIMAKOPOULOS, Nikolaos BARMPARESOS, Alexandros PANTAZARAS, Theoni KARLESSI and Siew Eang LEE. This chapter is based on a previous work by the authors, "On the comparison of occupancy in relation to energy consumption and indoor environmental quality: a case study", *Energy Procedia*, 134, 875–884, 2017, reproduced with permission from Elsevier.

For a color version of all figures in this chapter, see www.iste.co.uk/kolokotsa/smartbuildings2.zip.

still a subject of research. Recent research work has reported results on lighting, ventilation and air conditioning energy consumption in different climatic and cultural areas (Langevin *et al.* 2016). Both theoretical and experimental work indicates that for the Quebec and Rome climates, manual versus constant lighting control reduced lighting consumption by about 80% and that the cooling and primary power loads are significantly reduced by 42%, 57%, 60% and 43% respectively (Bourgeois *et al.* 2006). However, the main source of energy consumption of a building is due to the spread of the HVAC (Heating Ventilation and Air Conditioning) installations, which is increased by the growing demand for better thermal comfort. In fact, in developed countries, HVAC is the main energy consumption system of a typical dwelling since it is responsible for almost half of the total energy consumption, especially for the non-domestic cases (Yang *et al.* 2014). As far as the indoor environment is concerned, the increasing numbers of people in the building results in higher levels of carbon dioxide (CO_2), produced mainly by human exhalation (Vasile *et al.* 2016). It is well understood that CO_2 concentrations can be used as a tracer gas to evaluate the performance of the ventilation system (Seppänen 1999; Scheff *et al.* 2000). The aim of this work is to present the results of an experimental campaign that took place in an educational building of the National University of Singapore (NUS). The results refer to the energy consumption and air quality of three rooms, each with different usage and occupancy. The impact of different occupancy patterns on the energy demands, the illuminance of the building, as well as the internal levels of temperature, relative humidity and CO_2, were examined. The study of these parameters constitute an innovative contribution to the limited literature on energy consumption measurements within educational buildings of countries with hot and humid climates.

4.2. Methodology

The experimental campaign took place in three different types of rooms in terms of occupancy (an executive room, a computer room and a lecture theater), which were carefully selected from the block of buildings of the campus of the NUS. The main objective of the research was to investigate the impact of human presence in changing the energy consumption and indoor air quality within the building. In this respect, measurements were taken *in situ* on a continuous 24-hour basis for a period of four months. During the experiment, the daily uses of every room (lectures, meetings, presentations, etc.) and the operational details were also recorded. It should be noted that the rooms are located on different floors and that the building is situated in the city centre area. Energy consumption (kWh), illuminance (lux), temperature (°C), relative humidity (%) and CO_2 concentration (ppm) constitute the main measured parameters. Firstly, a raw database was created and analyzed qualitatively and quantitatively by statistical software tools. In order to achieve high quality data acquisition, Tukey's Two Sided test (Tukey 1949) was implemented to remove unwanted outliers and replace them with missing values. Subsequently, with the aid of the k-nearest neighbor (k-NN) imputation technique (Keerin *et al.* 2012), the missing data have been replaced by a weighted average of k nearest values. Thus, a smoother group of time series was achieved and unwanted measurements that could influence the final results were ignored. The results should initially give a general view of the three rooms' background for each measured parameter during the entire experimental period. Depending on this overview, the analysis focused on specific days of interest where particularly high levels of specific parameters would be further analyzed so the causes and conditions of this behavior could be investigated. Moreover, the study of any potential correlation between variables was facilitated so that final conclusions could be extracted. Finally, possible solutions and improvements were suggested where necessary.

4.3. Description of the case building

The building studied mainly contains offices, classrooms and a library space. The offices are open-plan, accommodating a large number of staff and researchers. The library is utilized by students and contains a few permanent staff members. Windows in air-conditioned spaces remain shut at all times during operating hours. The air conditioning system consists of multiple Air Handling Units (AHUs) and Fan Coil Units (FCUs), each serving separate areas. Some of the AHUs and FCUs are of the Single Coil Twin Fan (SCTF) system, which decouples the fresh and recirculated air streams until they are mixed before reaching the air-conditioned zone. The fresh air stream is regulated according to ventilation and indoor air quality requirements, based on readings from CO_2 concentration sensors in the return air grilles, while the recirculated air stream is regulated according to the cooling load, based on readings from temperature sensors. The plant consists of three chillers of which one is always on standby for extreme load conditions. Each chiller has its own dedicated set of pumps and cooling tower. Architectural plans, material properties, plant specifications and operating conditions were obtained from the building owner in the form of vendor documents, drawings and previous audit findings.

4.4. Description of the experimental procedure

Measurements were obtained by sensor kits, which contained four different types of sensors, measuring CO_2 concentration, temperature, relative humidity and illuminance. Table 4.1 demonstrates the sensors used as well as their technical specifications. Energy consumption data (per minute) was obtained from the building management system (BMS). Occupancy per room was provided by observation of researchers and by the number of Wi-Fi connections.

Environmental parameter	Sensor type	Range
Temperature (°C)	Sensirion SHT75	–40 to 123.8°C
RH (%)	Sensirion SHT75	0 to 100%
CO_2 (ppm)	K-30, CO_2 Meter Inc	0 to 5000 ppm
Illuminance (lux)	ROHM, B17xx	0 to 65000 lux

Table 4.1. *Technical specifications of the sensors*

One sensor is placed in the executive office, two in the computer room and three in the lecture theater. Greater interest was shown in rooms with frequent fluctuation in occupancy such as the lecture theater and the computer room. Data was collected at one minute sampling intervals.

4.5. Results

4.5.1. *Investigation of energy consumption and indoor air quality*

The main objective of this work was to investigate the indoor environmental quality and the energy consumption of three rooms. As a first step, it was important to study their energy and environmental behaviour. Table 4.2 summarizes monthly values of the measured parameters for all three experimental sites, thus allowing a total overview of the obtained results. Figure 4.1 and Figure 4.2 show the total energy consumption and the total luminous emittance in the three rooms. It can be seen that the computer room and the lecture theater present the highest energy consumption, as is expected, since they are rooms with high occupancy rates but also with equipment (presentation equipment and computers). The executive room, which is not used very often, demonstrates low values of energy consumption and illuminance most of the time, as expected. Seasonality seems to be a key factor, since measurements during August and September presented much higher levels than their respective measurements in October and November.

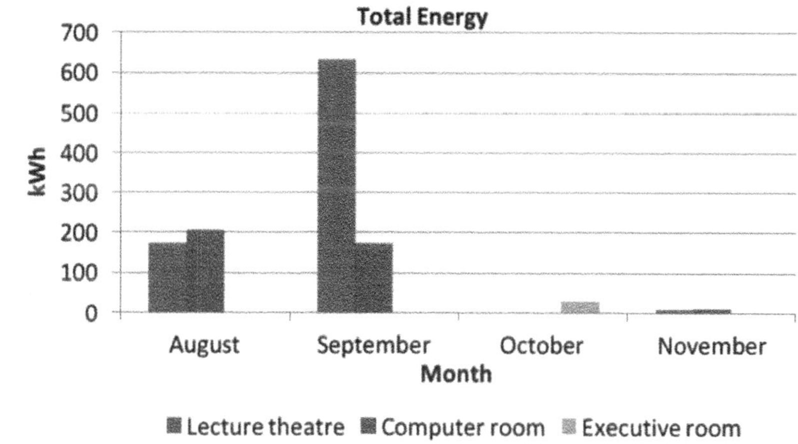

Figure 4.1. *Total energy (excluding lighting) for the three rooms*

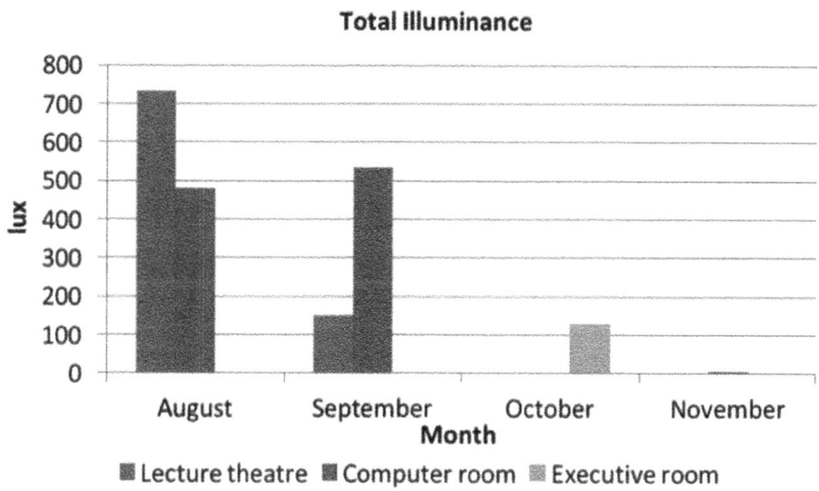

Figure 4.2. *Total illuminance for the three rooms*

The temperature and relative humidity measurements are illustrated in two different types of graphs. Firstly, in order to assess the overall thermal behaviour of the rooms, the box plot approach was used. Figure 4.3 presents the box plots of the (a) temperature and (b) relative humidity. We can see that the lecture theater, which has the highest variability in terms of occupancy, shows the lowest mean temperature and the highest variation of temperature values. This is expected since the lecture theater is not always booked for lectures and, even if it is, the number of people varies depending on the lecture topic.

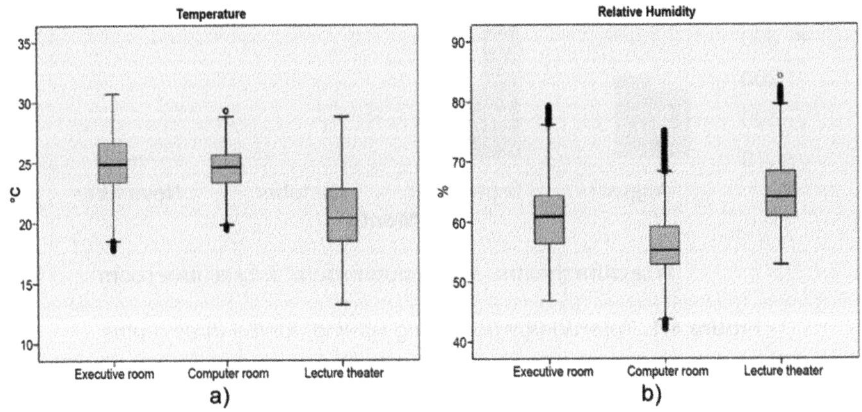

Figure 4.3. *Box plots for a) temperature and b) relative humidity of the three rooms*

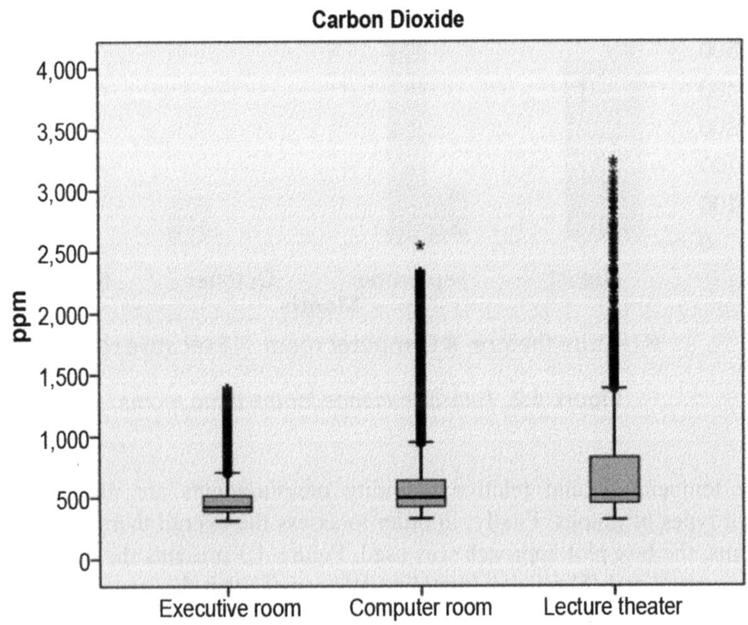

Figure 4.4. *CO_2 concentrations in the three rooms*

Figure 4.4 shows the CO_2 concentrations in the three rooms. Once again, the lecture theater presents greater variability as a result of its varying occupancy patterns. From the above graphs, it is easy to conclude that the lecture theater presents the most interesting environmental behavior, due to the complexity of its occupancy patterns.

Experimental sites	**Parameters**	**Year 2016**			
		August	**September**	**October**	**November**
Lecture Theater (sensors 70, 5, 13)	Total Illuminance [=lux]	732.5	150.3	0.0	3.0
	Total Energy [=kWh]	173.7	632.7	0.0	8.6
	Average Temperature [=°C]	20.9	20.1	21.0	21.4
	Average Relative Humidity [=%]	67.5	65.2	63.0	65.8
	Average CO_2 [=ppm]	817.0	635.7	705.0	543.5
Computer Room (sensors 50, 37)	Total Illuminance [=lux]	480.6	535.0	0.0	4.9
	Total Energy [=kWh]	205.6	173.6	0.0	9.7
	Average Temperature [=°C]	24.8	24.7	24.6	24.1
	Average Relative Humidity [=%]	58.3	56.1	54.8	57
	Average CO_2 [=ppm]	554.2	559.3	655.8	523.6
Executive Room (sensor 20)	Total Illuminance [=lux]	0.0	0.0	126.1	0.0
	Total Energy [=kWh]	0.0	0.0	27.4	0.0
	Average Temperature [=°C]	25.2	25	24.8	24.4
	Average Relative Humidity [=%]	60.7	59.8	59.9	61.7
	Average CO_2 [=ppm]	454.1	464.9	513.7	472.6

Table 4.2. *Monthly values for all measured parameters within the three rooms*

4.5.2. *Days of special interest – high occupancy*

A specific day with high occupancy was chosen to show the daily variation of temperature, relative humidity and CO_2 concentration, as well as the fluctuation of illuminance and energy consumption in accordance with the number of people in the lecture theater (Figures 4.5, 4.6 and 4.7). Temperature and relative humidity were completely controlled by the air conditioning system and remained relatively stable. However, at the beginning of a lecture (and during the class) when the theater is almost full, CO_2 is elevated to rather alarming levels (higher than the 1,000 ppm ASHRAE limit). Moreover, temperature also rises, but it does not exceed 25°C. As for energy consumption, it seems to be particularly influenced by the presence of people within the theater, as the peaks follow almost the same pattern as the occupancy. On the other hand, illuminance does not seem to be strongly influenced by human presence. It is found to be elevated during the morning hours, with a small number of people inside and reduced in the early afternoon, with much more occupancy during class. Thus, the levels of internal lighting mainly depend on the openings of the room and human activities (such as turning off the lights or using blinds or curtains) rather than on the number of people inside the lecture theater. This conclusion is in agreement with the research of (Mahdavi *et al.* 2008), who observed that there is a certain relationship but not a strict correlation between occupancy and (electrical) lighting operation.

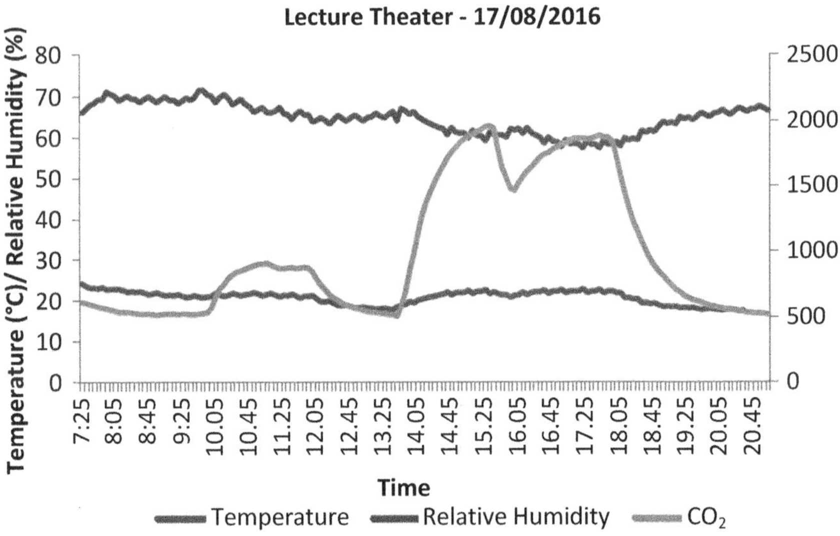

Figure 4.5. *Lecture theater on a day with high occupancy*

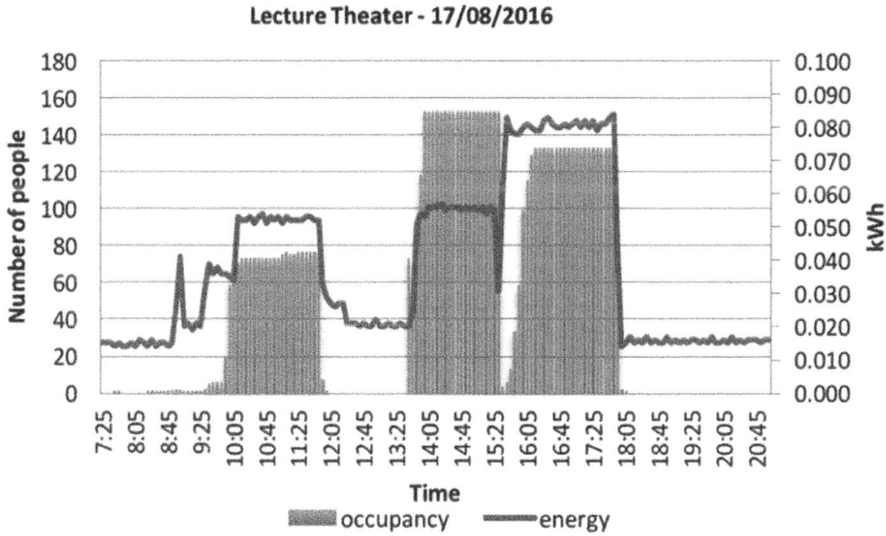

Figure 4.6. *Energy consumption levels in lecture theater with respect to occupancy*

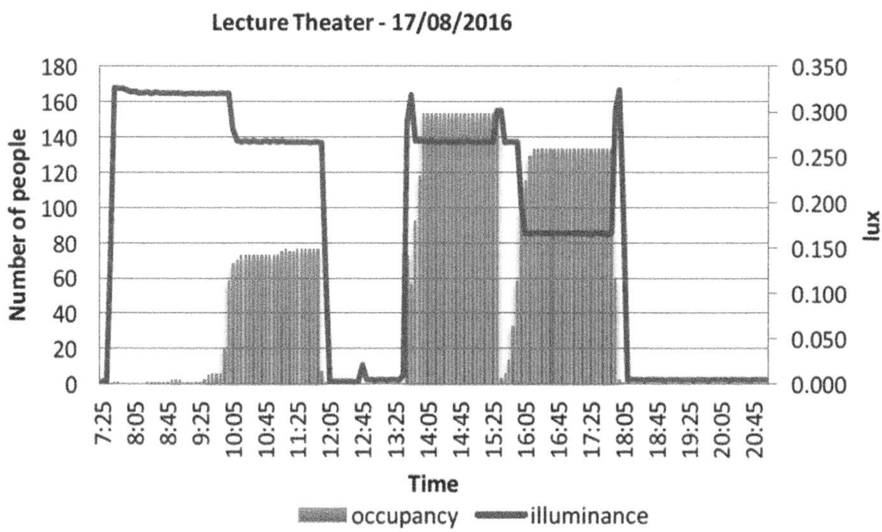

Figure 4.7. *Illuminance levels in lecture theater with respect to occupancy*

4.5.3. *Days of special interest – increased energy consumption*

Another case study of interest was based on the analysis of a specific day in September when high levels of energy consumption were observed within the computer room. Figure 4.8 shows a correlation between energy consumption and the levels of CO_2 with a coefficient of determination (R2) equal to 0.76. The result is statistically significant, as the p value of the ANOVA test approached zero. In this case, the concentration of CO_2 was used as an indicator of occupancy inside the room and demonstrated that the presence of people influences energy values mainly because of the electric load produced by the computers. The same result was observed in the experiment of (Hestnes and Kofoed 2002) where the electricity consumption presented a significant positive correlation with the occupancy rate, even with a large proportion remaining constant over time. We note that temperature, relative humidity and illuminance remained stable in the room with no significant fluctuations.

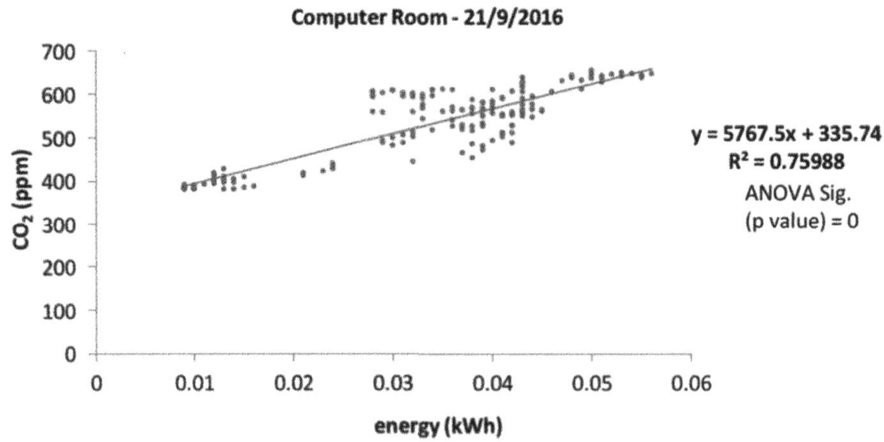

Figure 4.8. *Scatter plot of CO_2 levels vs energy consumption for a specific day in the computer room*

4.6. Discussion and concluding remarks

The number of people and their activities within a building, as well as its general structure (dwellings, offices, hospitals, educational institutions, etc.) are the main factors influencing the energy consumption and the indoor air quality. A key question is whether the number of people affects each of the examined variables and how the optimal use of the building is ensured from the energy point of view. For that reason, three rooms with different characteristics, a lecture theater, an executive office and a computer room were selected as experimental sites. The lecture theater presented a high variability for all parameters as it was not occupied frequently, and high levels of energy demands in September were recorded. In the computer room, high values of illuminance were recorded during August and September. For the executive room, low values of illuminance and energy consumption were measured only in October. It is clear that there is a strong correlation between energy consumption and occupancy. The levels of illuminance do not seem to be strongly influenced by the amount of people in each room but mainly from human activity. Indoor concentration of CO_2, which is related to human exhalation, was found to be elevated during class hours while temperature and relative humidity remained relatively stable because of the existing air conditioning system.

The results can be summarized as follows:

– occupancy is strongly correlated only with energy consumption and CO_2 concentrations;

– illumination is mainly influenced by the building's characteristics and human activities but not by occupancy levels. During a lecture, a large number of people may be in a relatively dark environment while under certain circumstances (exam periods or studying), a small amount of people may turn on all the lights;

– because of the presence of an HVAC system, temperature and relative humidity remained unaffected by human presence and remained relatively stable for all three rooms;

– the room with the largest fluctuations of all variables was the lecture theater as it was used greatly at regular time intervals, depending on the schedule. Occupancy affects specific parameters (energy and CO_2) within an educational building of a hot and humid climate area. Since the number of people that were using the specific section of the building cannot be changed, smart occupancy sensors could be proved an appropriate solution. Similar research (Garg and Bansal 2000) indicates that these sensors can save up to 5% more energy. This can prove useful for the present case of building. Reduction of CO_2 levels can be achieved through the improvement of the operation conditions of the existing ventilation system (mechanical or natural). Specific research (Pantazaras *et al.* 2016) can help in this direction as it demonstrates predictive models of CO_2 levels for various types of buildings. With respect to future work, further research can be conducted into two different directions. Firstly, as seasonality proved to play an important role in the final results, it is proposed that experiments are continued throughout an academic year in order to present a comprehensive annual behaviour of variables. Secondly, it would be of great interest to carry out similar measurements after the implementation of the proposed improvement measures, in order to ascertain the percentage of energy savings in the building and the decrease in CO_2 levels.

5

Indoor Environmental Quality and Energy Consumption Assessment and ANN Predictions for an Integrated Internet-based Energy Management System Towards a Zero-energy Building

5.1. Introduction

Universities' campuses and campus buildings can be viewed as small districts and microgrids. Therefore, the School of Design and Environment (SDE) at the National University of Singapore intends to integrate all buildings under a common energy and indoor environmental quality management platform. Moreover, all SDE buildings are currently under refurbishment while a new building (SDE 4) is under construction.

The overall procedure requires the collection of data on the energy consumption as well as on the indoor environmental quality. The monitoring activities for all the NUS SDE buildings, with emphasis on the SDE 3 building, are included in section 5.3. The energy consumption and indoor environmental quality are evaluated and cross correlated. Prediction algorithms are tested using artificial neural networks. The specifications for the integrated internet-based installation for SDE is included in section 5.7, while the conclusions and future research can be found in section 5.8.

Chapter written by Denia KOLOKOTSA.

For a color version of all figures in this chapter, see www.iste.co.uk/kolokotsa/smartbuildings2.zip.

5.2. Description of the SDE buildings

5.2.1. *General information*

The building under study (SDE 3) is one of the three buildings of the School of Design and Environment of the National University of Singapore. The building's location is pinpointed in Figure 5.1, and the other SDE buildings (SDE 1 and SDE 2), as well as the site for the SDE building under construction, are also depicted.

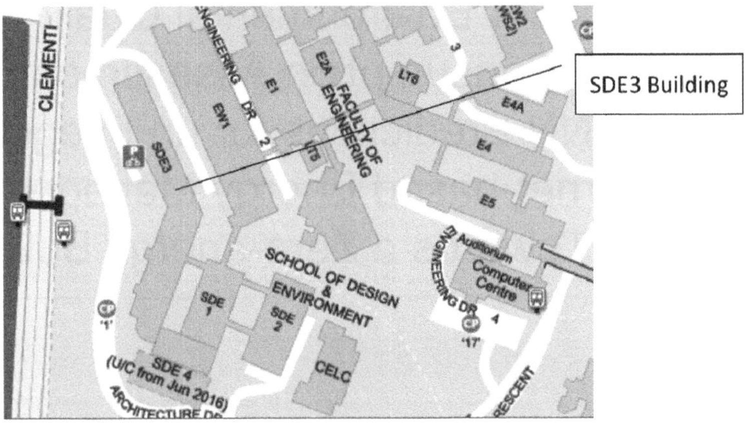

Figure 5.1. *The SDE 3 building*

Pictures of the SDE 3 building are shown in Figure 5.2.

Figure 5.2. *The entrance and front elevation of SDE 3*

Climate zone	Tropics
Year of construction	1970
Building utilization i.e. residential, office, commercial, industrial, etc.	Offices, laboratories and classrooms
Building occupancy schedule	08:00–19:00
Gross floor area (m^2)	8680.7 m^2 (7946.7 conditioned and 734 non-conditioned)
Number of storeys	Four
Architectural plans or 3D images of the building	Presented in section 5.3
Building envelope description (i.e. wall – roof – floor structure, type of façade, percentage of window to wall area, roof type, etc)	External wall: 50 mm gypsum and concrete and 25 mm gypsum board Floor: 100 mm lightweight concrete and 150 mm concrete Interior wall: 25 mm gypsum board and wall insulation and 25 mm gypsum board
HVAC system (kW)	The building is serviced by FCU systems with thermostats available in each room 3*14.92 kW and 1*7.46kW VRV inverter heat pumps
Lighting system (kW)	Fluorescent lamps
Current state of repair and year of last major renovation of building envelope and main technical installations, respectively	The building is fairly maintained but has not been upgraded significantly in terms of energy performance during recent years
Peak load (kW)	1250 kW

Table 5.1. *Building envelope and systems of SDE 3*

5.2.2. *Monitoring activities for SDE 3*

In the SDE 3 building a number of sensors and smart meters are installed. The monitoring equipment is tabulated in Table 5.2. The sensors' position around the building spaces and floors is depicted in Figures 5.3–5.6.

Environmental parameter	Sensor	Specifications		
		Range	Accuracy	Resolution
Temperature and relative humidity	Sensirion SHT75	–40 to 123.8°C	±0.3°C (@25°C)	0.01°C
		0 to 100% RH	±1.8% RH	0.03% RH
CO_2	K-30, CO_2 Meter Inc.	0 to 5,000 ppm	±30 ppm + 3% of measured value	–
Illuminance	ROHM B17xx series	0 to 65,000 lx	–	1 lx

Table 5.2. *The monitoring equipment of the SDE 3 building*

5.3. The power loads and hourly energy consumption

Measuring the building's energy consumption started in April 2016. The hourly power loads during the last period (i.e. April and May 2016) are depicted in Figure 5.7.

The histograms of energy consumption for the whole period and for the two months of April and May 2016 are depicted in Figure 5.8 and Figure 5.9 respectively. The mean value is 280kW for the whole period. We can see that the peak loads are significantly reduced during May. The total energy consumption for April is 235,626 kWh while for May it is reduced to 176,010 kWh, which corresponds to *almost 25% less* energy than April.

5.4. Indoor environmental quality

The IEQ assessment is performed based on the data collected from the various sensors.

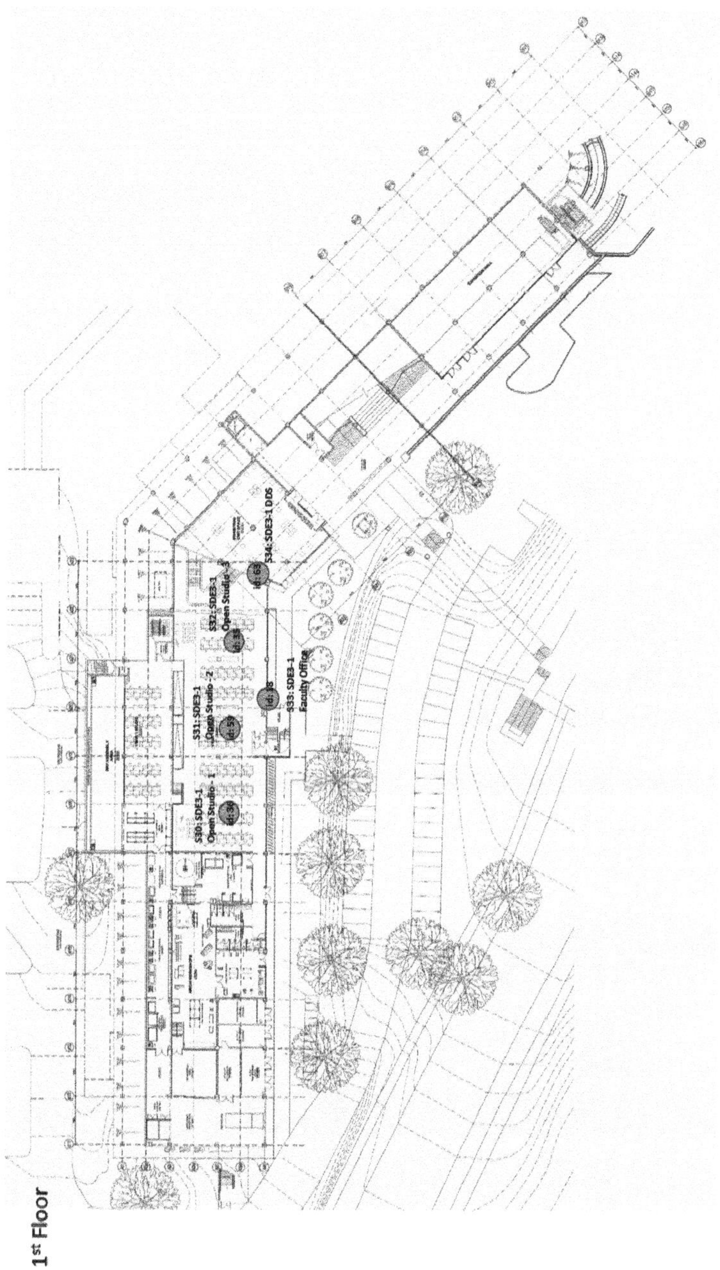

Figure 5.3. *The position of sensors on the 1st floor of SDE 3*

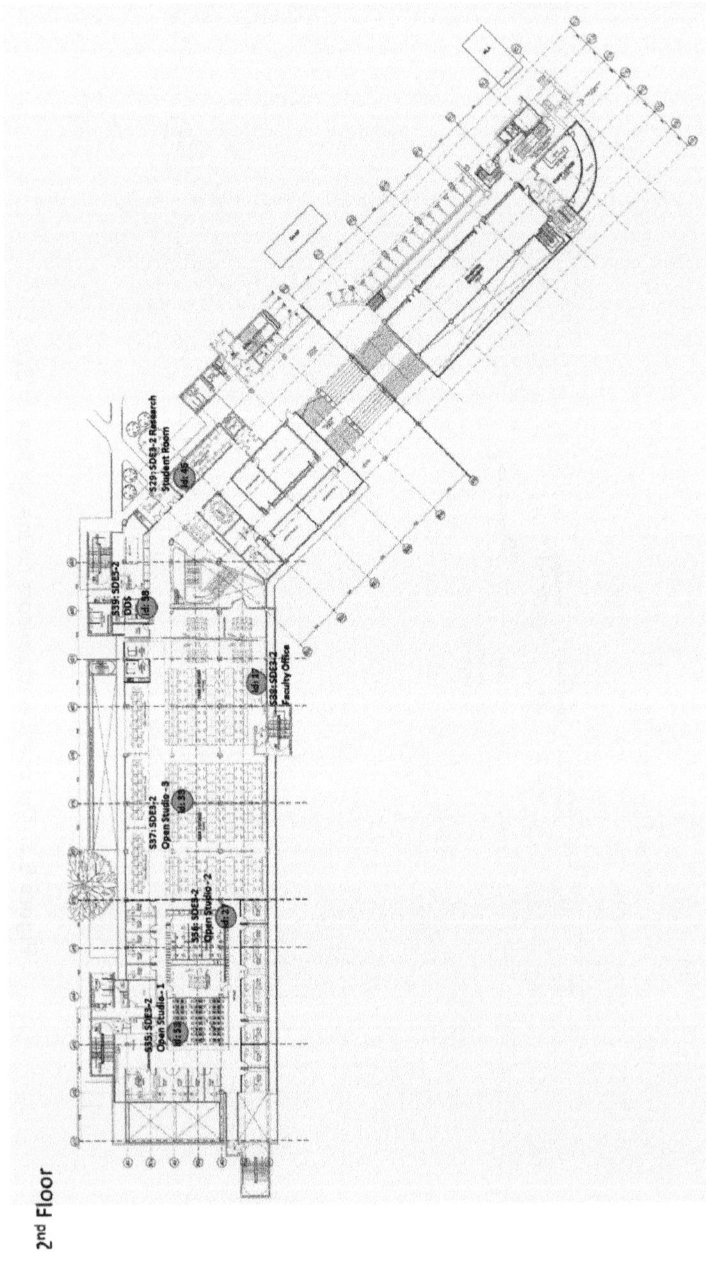

Figure 5.4. *The position of sensors on the 2nd floor of SDE 3*

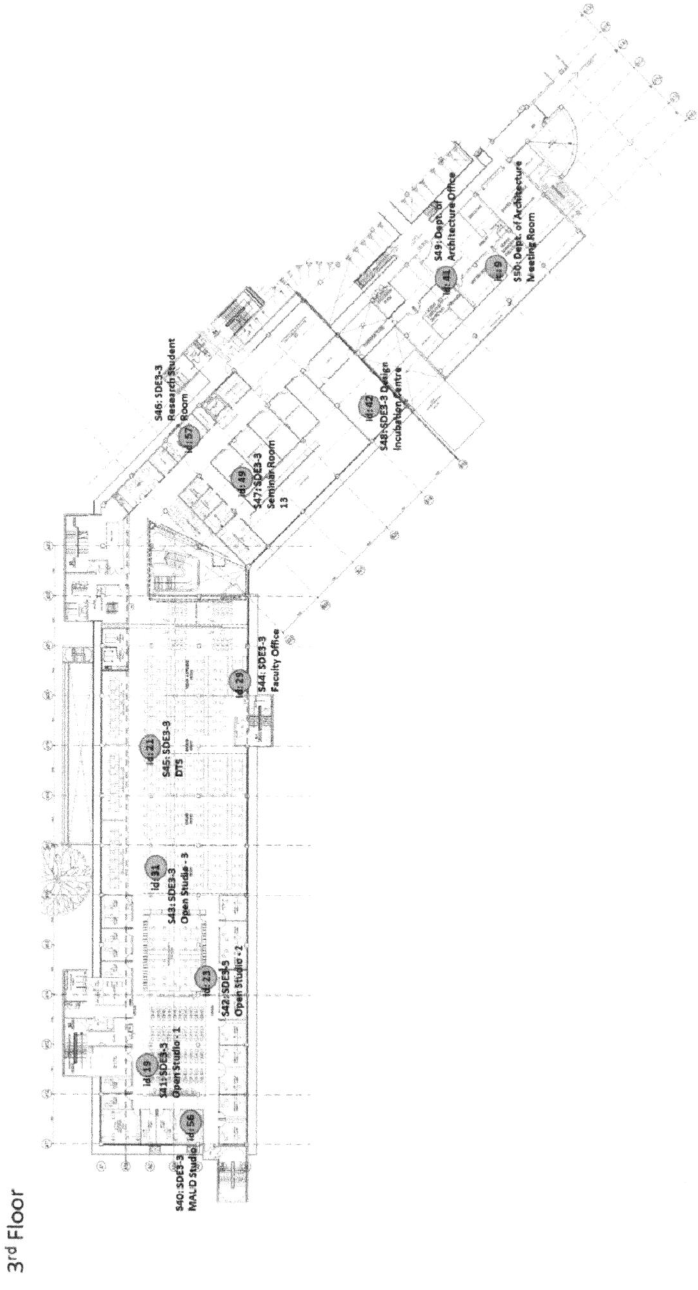

Figure 5.5. *The position of sensors on the 3rd floor of SDE 3*

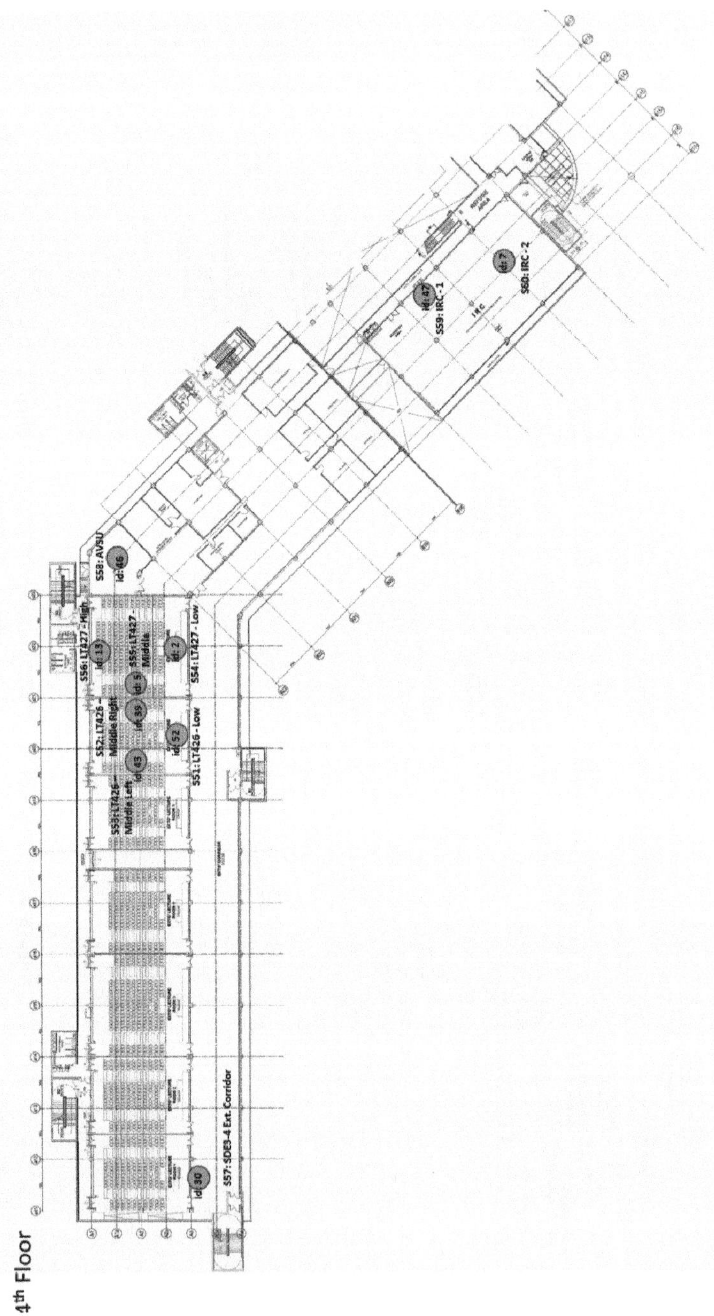

Figure 5.6. *The position of sensors on the 4th floor of SDE 3*

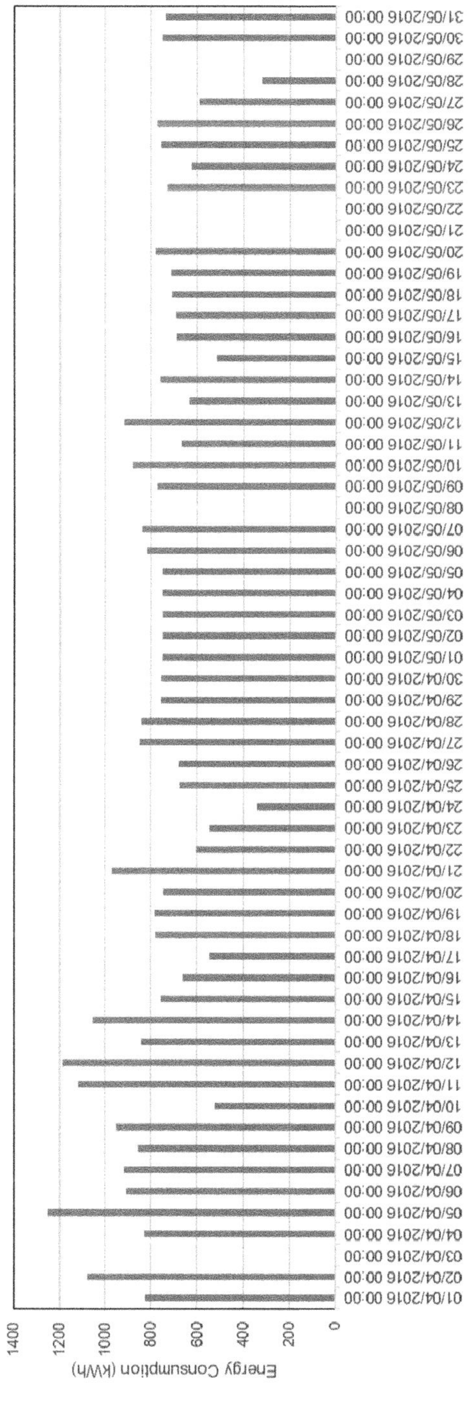

Figure 5.7. *The energy consumption of the SDE 3 building during April and May 2016*

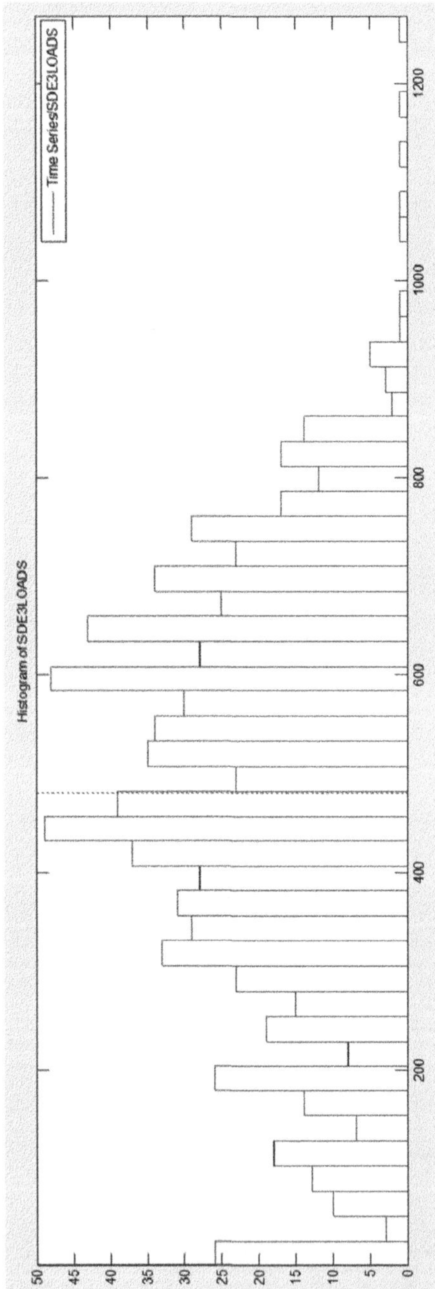

Figure 5.8. *The power loads histogram (excluding zero values)*

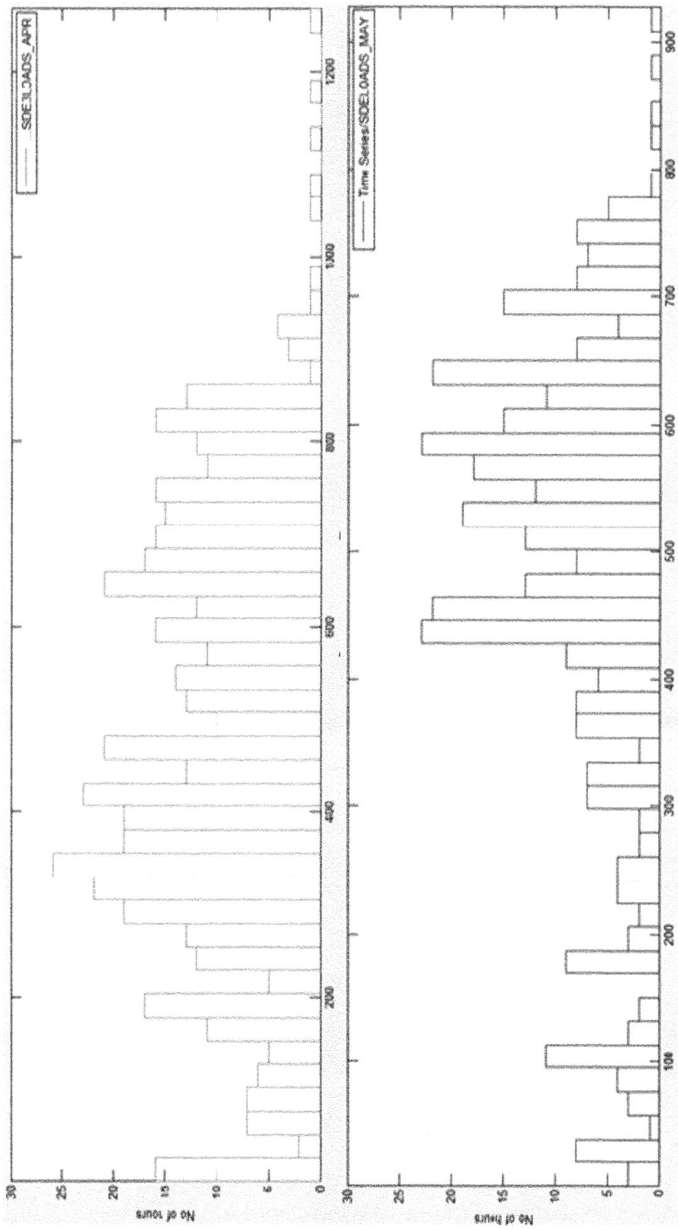

Figure 5.9. *The energy loads per month for April (top) and May (bottom) 2016*

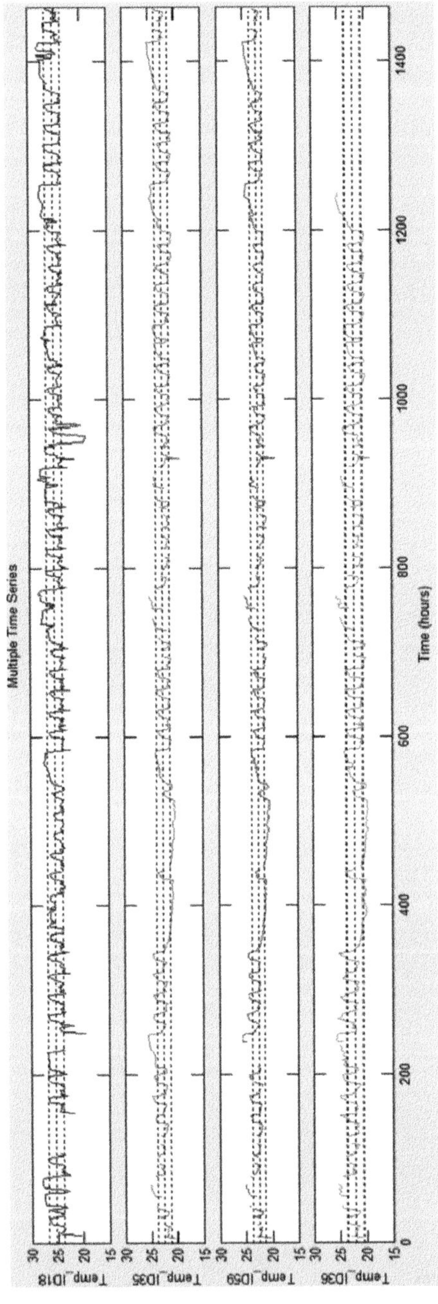

Figure 5.10. *The air temperature of 1st floor time series*

5.4.1. *Thermal comfort assessment – time series analysis*

A time series analysis is performed for the data collected. Indicatively, the time series of the 1st floor sensors (i.e. ID_18, 35, 36 and 59) are depicted in Figure 5.10. We can see that for room ID_18, the temperature reaches quite high values.

The air temperature statistics for the temperature data are tabulated in Table 5.3. We can see that for the 1st floor the indoor air temperature is lower than for the other floors. It is assumed that the air conditioner is in full operation all day long for the first floor rooms apart from in room 18. For the 2nd, 3rd and 4th floors, the indoor temperature reaches 28–30°C during the time that the air conditioner is turned off.

Concerning relative humidity for the indoor environment, the values at the various points range within the comfort zone, as depicted in Figure 5.11. The outdoor relative humidity values are very high and are considered to be a very significant parameter for the indoor thermal comfort and the energy consumption for air conditioning.

Sensor	Mean (°C)	Median (°C)	STD (°C)	Tmin (°C)	Tmax (°C)
1st FLOOR					
ID_18	25.32	25.32	1.47	19.07	28.48
ID_35	22.52	22.43	1.31	19.31	25.65
ID_36	22.13	21.91	1.58	19.10	25.71
ID_59	22.45	22.43	1.14	19.14	25.63
2nd FLOOR					
ID_17	26.2	26.2	1.2	23.5	29.3
ID_27	25.1	25.0	1.8	21.3	29.4
ID_45	25.8	26.1	1.8	22.1	29.8
ID_58	24.2	24.1	1.8	20.1	31.2
3rd FLOOR					
ID_19	24.7	24.5	1.8	21.4	29.2
ID_31	24.8	24.7	1.9	21.2	29.7
ID_42	26.0	25.7	2.2	21.2	31.3
ID_49	23.0	24.0	2.7	17.7	27.3
4th FLOOR					
ID_13	24.6	25.7	3.1	17.2	30.1
ID_43	24.9	25.8	2.9	18.1	28.7
ID_46	26.2	26.7	1.9	21.5	29.4

Table 5.3. *The air temperature statistical analysis*

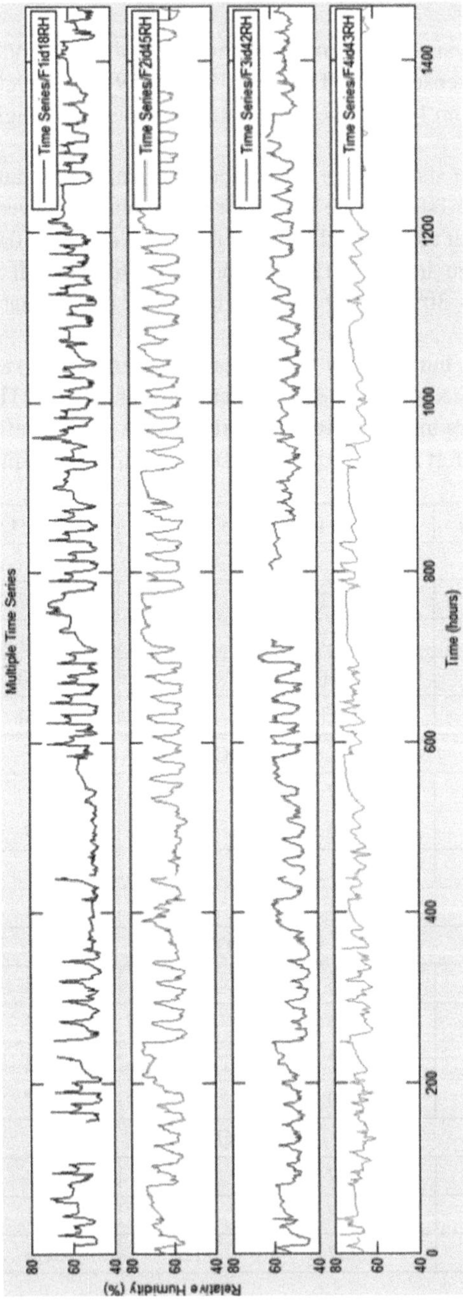

Figure 5.11. *The relative humidity time series*

5.4.2. *Indoor air quality*

The indoor air quality is assessed via the measurement of the CO_2 levels. The statistics of the 1st floor data are tabulated in Table 5.4 while the time series are depicted in Figure 5.12. Apart from a specific period in April where high CO_2 concentration occurs, for the rest of the period the levels of CO_2 remain within the acceptable limits.

Sensor	Mean (ppm)	Median (ppm)	STD (ppm)	CO_2_min (ppm)	CO_2_max (ppm)
ID_18	474.25	458.5	71.65	371.03	907.5
ID_35	423.27	385.53	101.88	310	963.5
ID_36	403.35	358.5	114.7	272.9	975.5
ID_59	459.6	428.7	92.3	353.3	1,031

Table 5.4. *The statistical analysis for CO_2 concentration*

5.4.3. *The indoor illuminance levels*

The indoor illuminance is assessed via the measurement of the illuminance levels. The statistics of the 1st floor data are tabulated in Table 5.5 while the time series are depicted in Figure 5.13. We can see that the illuminance level is within 0–300 lux for most rooms, but greater fluctuations occur in the room containing sensor ID_18.

Sensor	Mean (lx)	Median (lx)	STD (lx)	ILL_min (lx)	ILL_max (lx)
ID_18	203.2	4.09	333.9	0	976.8
ID_35	159.4	174.5	59.9	7.5	248.2
ID_36	191.1	255.2	102.2	3.4	288.4
ID_59	129.9	134.8	42.8	12.5	213.6

Table 5.5. *The statistical analysis for illuminance levels on 1st floor*

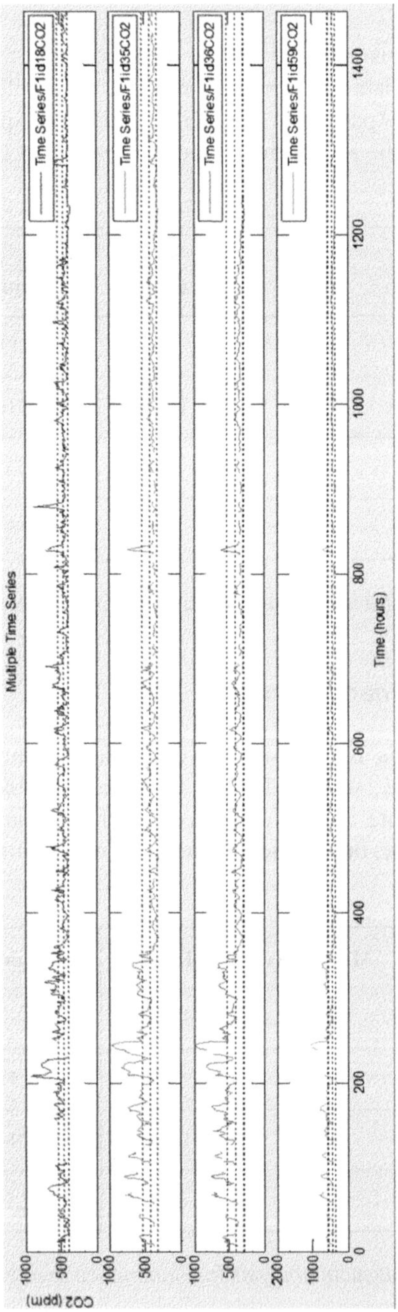

Figure 5.12. *The time series of CO_2 concentration measurements*

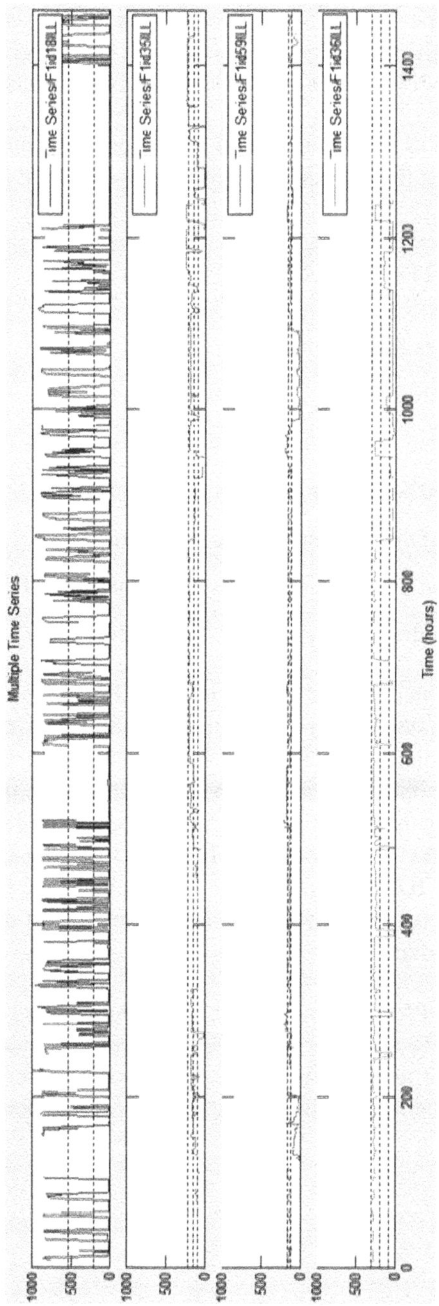

Figure 5.13. *The time series of illuminance measurements (lux)*

Concerning the thermal comfort of the SDE 3 building, three more indices are examined. The first is the discomfort index (DI), as it combines the air temperature and relative humidity (see equation [5.1]) (Tselepidaki 1992) and the others are the PMV and PPD indices (ISO 2005; Katafygiotou and Serghides 2014).

The DI is expressed as a combination of air temperature and relative humidity. The SDE 3 building is in the tropics where the relative humidity plays a significant role. The DI is:

$$DI = T_a - 0.55(1 - 0.01 \cdot RH) \cdot (T_a - 14.5) \qquad [5.1]$$

The various categories for the DI are:

DI < 21 – No discomfort

DI = 21–24 – Less than 50% of the population feels discomfort

DI = 25–27 – Over 50% of the population feels discomfort

DI = 28–29 – Most of the population feels discomfort

DI =30–32 – The whole population feels significant discomfort

DI > 32 – State of medical emergency

Although the DI is mainly used for outdoor conditions, it is calculated for the SDE building since it combines two of the thermal comfort variables measured. The results of the DI calculation are included in the boxplots of Figure 5.14 and Figure 5.15 for the first floor and all floors, respectively. We see that there are some variations, especially for the ID_18 and ID_59 rooms, for the first floor. The room with the lowest DI is ID_36. The highest DI occurs for the ID_18 room on the 1st floor, as well as the ID_45 room, the ID_42 room and the ID_43 room for the 2nd, 3rd and 4th floors, respectively. The time that the highest discomfort occurs is during non-occupancy hours.

For these rooms, the PMV for one typical day of April is calculated and depicted in Figures 5.16, 5.17, 5.18 and 5.19 (ISO 2005). A sensitivity analysis is performed for various mean radiant temperature values, since no measurements are available. For the specific rooms the PPD is always less than 5% during occupancy hours.

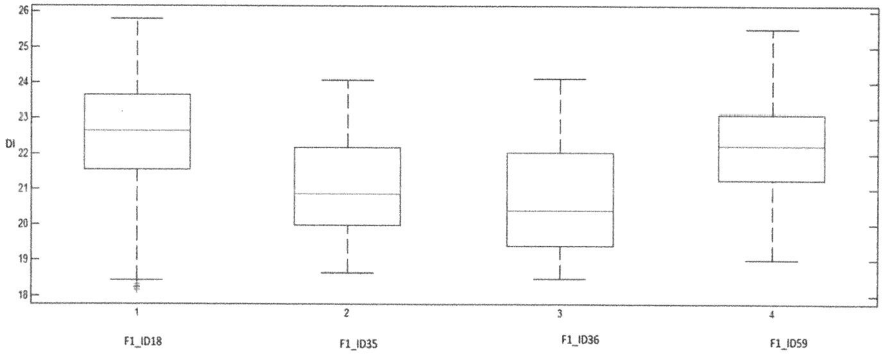

Figure 5.14. *The discomfort index for the 1st floor*

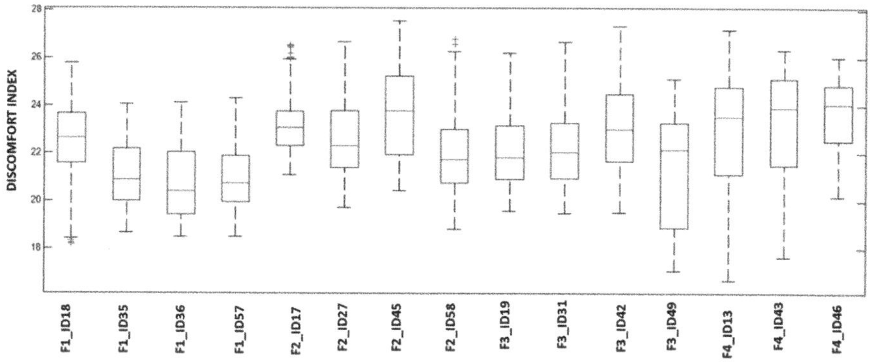

Figure 5.15. *The discomfort index for all floors*

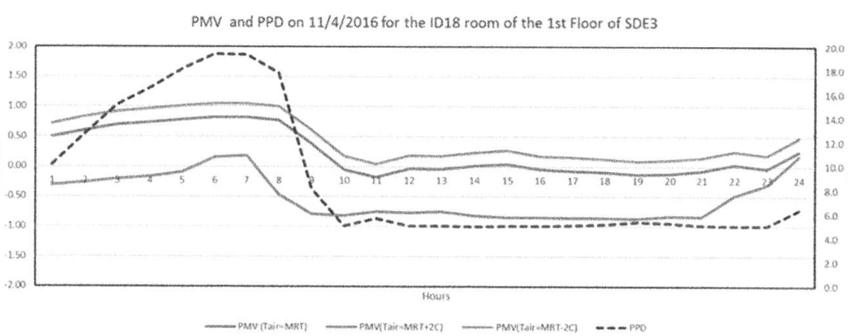

Figure 5.16. *The PMV and PPD index of the ID_18 room on April 11, 2016*

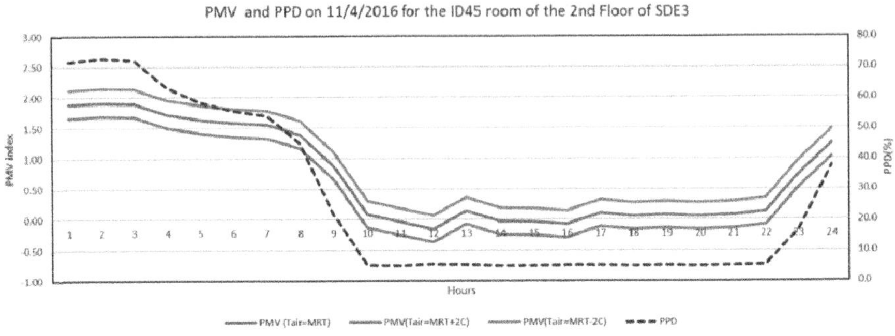

Figure 5.17. *The PMV and PPD index of the ID_45 room on April 11, 2016*

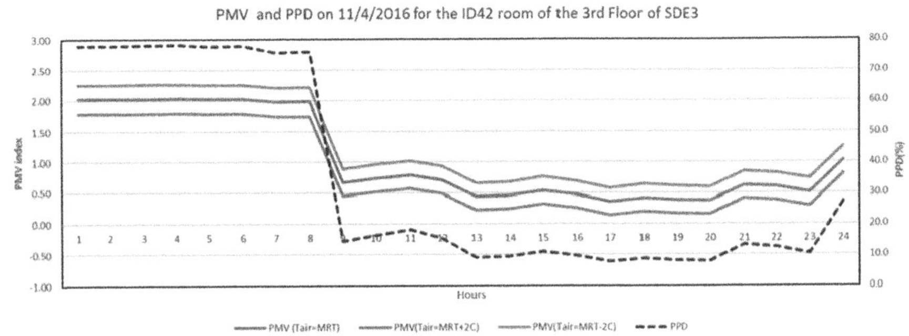

Figure 5.18. *The PMV and PPD index of the ID_42 room on April 11, 2016*

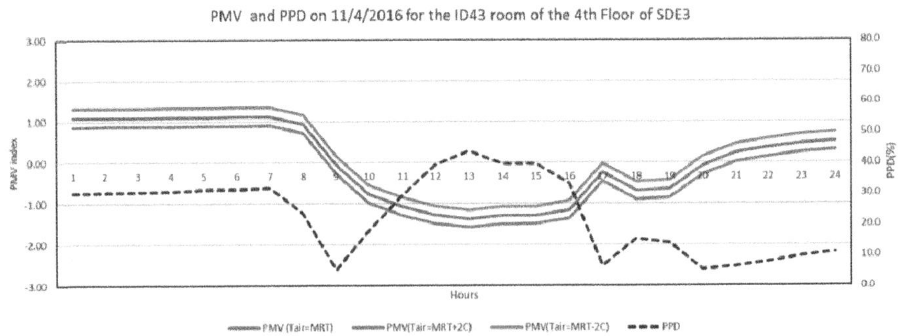

Figure 5.19. *The PMV and PPD index of the ID_43 room on April 11, 2016*

5.5. Cross correlation

A cross correlation is performed between the air temperature measured at the various points and the loads as well as air temperature. The indoor temperature and relative humidity versus the power loads have the same pattern, as depicted in Figures 5.20 and 5.21, respectively, while the cross correlation between relative humidity and temperature has a completely different pattern (Figure 5.22).

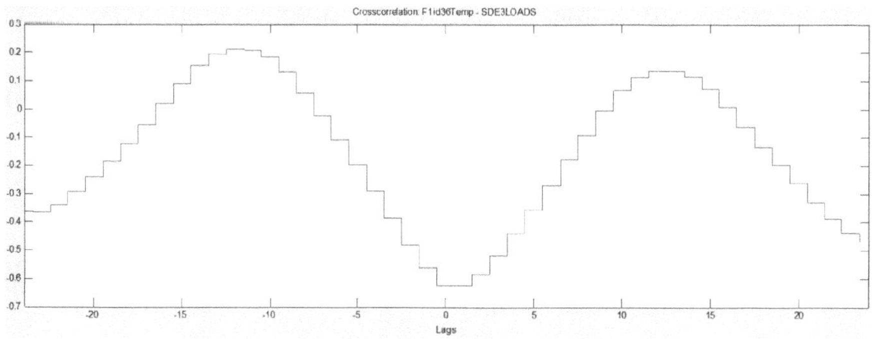

Figure 5.20. *Cross correlation between indoor air temperature and power loads for the 1st floor*

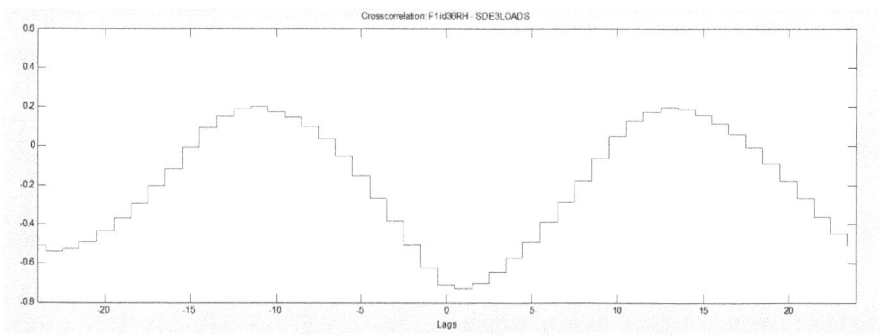

Figure 5.21. *Cross correlation between relative humidity and power loads for the 1st floor*

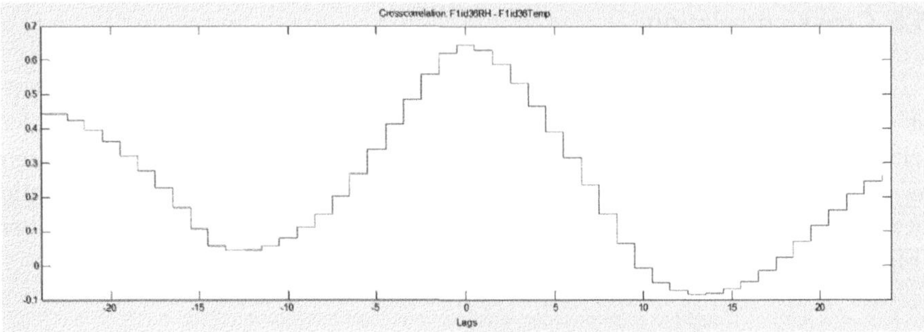

Figure 5.22. *Cross correlation between relative humidity and temperature for the 1st floor*

5.6. Prediction using artificial neural networks (ANN)

Various types of artificial neural network (ANN) have been developed and tested for the prediction of power loads as well as weather conditions (Gobakis *et al.* 2011; Joaquim and Rosa 2005).

An ANN is an information processing paradigm that is inspired by the way biological nervous systems such as the brain process information. The key element of this paradigm is the novel structure of the information processing system. It is composed of a large number of highly interconnected processing elements (neurons) working in unison to solve specific problems. ANNs, like people, learn by example. The central processing component of a neural network is a neuron, which can process a local memory and carry out localized information. Each neuron computes a weighted sum of the inputs it receives and adds a bias *(b)* to form the net input *(x)*. The bias is included in the neurons to allow the activation function to be offset from zero:

$$x = w_{1,1} \cdot p_1 + w_{1,2} \cdot p_2 + \ldots + w_{1,j} \cdot p_j = b \qquad [5.2]$$

The net input *(x)* is then passed to the subsequent layer through a non-linear sigmoid function to form its own output:

$$y_j = \frac{1}{1+e^{-x}}$$

Afterwards, the output y_j is compared with the target output t_j using an error function of the form:

$$\delta_K = (t_j - y_j) y_j (1 - y_j)$$

For the neuron in the hidden layer, the error term is given by the following equation:

$$\delta_j = y_j (1 - y_j) \sum \delta_K w_k$$

where δ_K is the error term of the output layer and w_K is the weight between the hidden layer and output layer (Elminir 2007).

For the SDE 3 building, we considered that the prediction of the outdoor relative humidity together with the outdoor air temperature can provide some insights on the power loads required to maintain indoor comfort. The nonlinear autoregressive network with exogenous inputs (NARX) is used for this specific case. NARX is a recurrent dynamic network, with feedback connections enclosing several layers of the network. The NARX model is based on the linear ARX model, which is commonly used in time-series modeling. The Levenberg–Marquardt training algorithm is used.

5.6.1. *Prediction of outdoor temperature*

The input data for each ANN are:

– year, month and day;

– pressure (kPa);

– solar radiation (W/m^2).

The output is the outdoor temperature for various prediction horizons. For 24 hours ahead prediction we have R = 0.94 which is quite promising (see Figure 5.23).

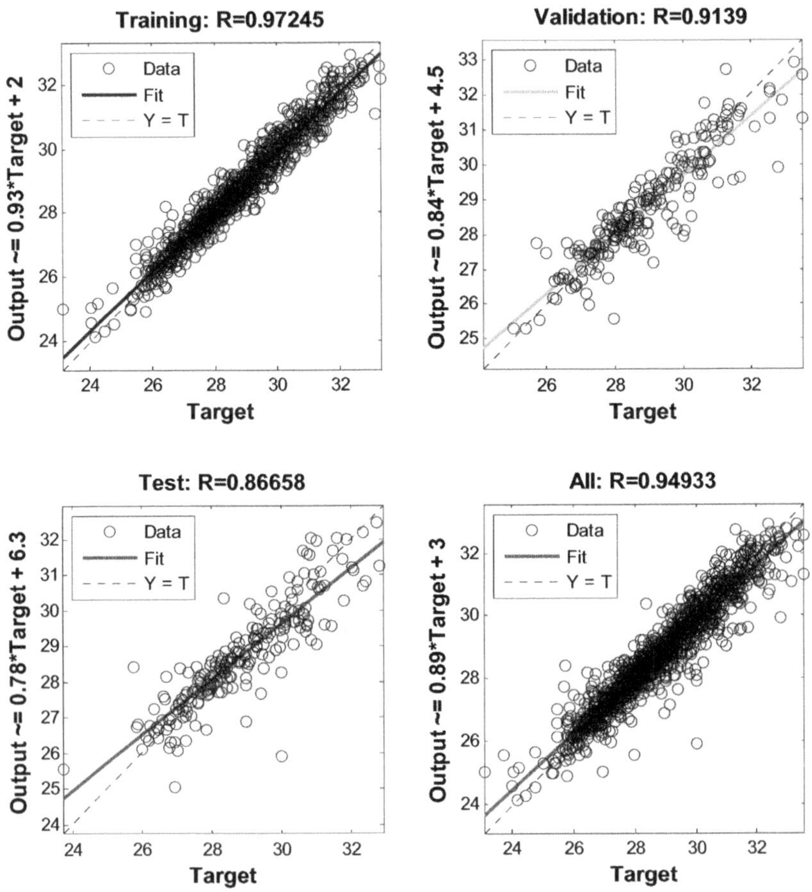

Figure 5.23. *The prediction of outdoor temperature 24 hours ahead using NARX*

5.6.2. *Prediction of relative humidity*

The input data for each ANN are:

– year, month and day;

– pressure (kPa);

– solar radiation (W/m^2).

The output of the ANN is the outdoor relative humidity for the 24 hours prediction horizon (see Figure 5.24).

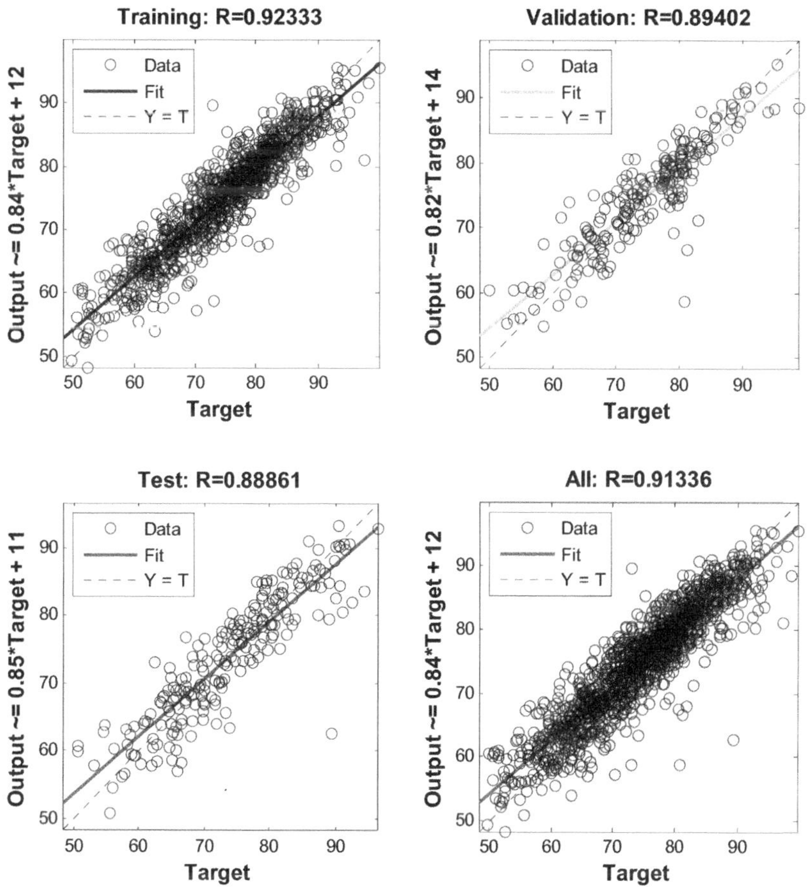

Figure 5.24. *The prediction of outdoor relative humidity 24 hours ahead*

5.6.3. *Prediction of power loads*

The prediction of loads 24 hours ahead using NARX is depicted in Figure 5.25 and Figure 5.26. The cases examined are:

Case 1: the input data used for this specific case (case 1) are:

– year, month and day;

– pressure (kPa);

– solar radiation (W/m^2).

Case 2: the input data are:

– year, month and day;

– pressure (kPa);

– solar radiation (W/m^2);

– air temperature;

– outdoor relative humidity.

We see that the increase in the number of inputs does not considerably influence the results.

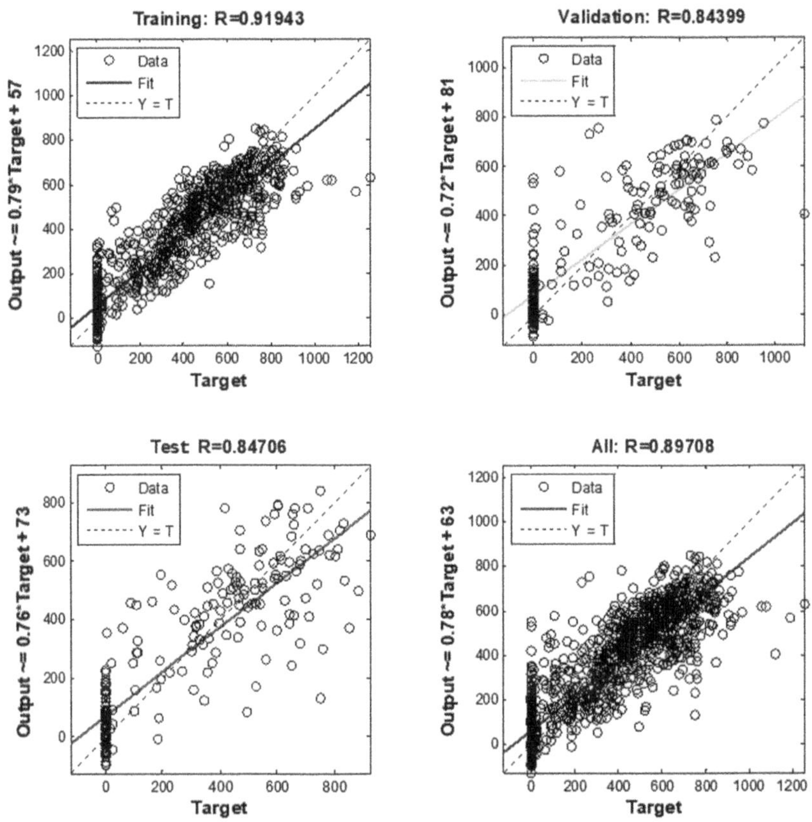

Figure 5.25. *The prediction of power loads 24 hours ahead for case 1*

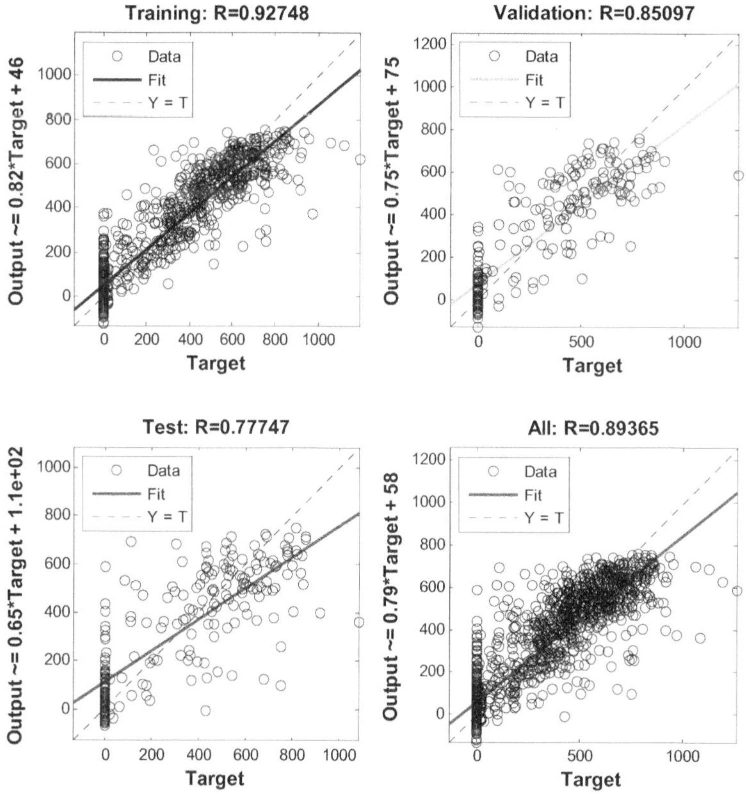

Figure 5.26. *The prediction of power loads for case 2*

5.7. Specifications for an integrated internet-based energy management system towards a zero-energy building

The scope of the integrated internet-based energy management for SDE

During recent decades the role of information and computer technology (ICT) for energy efficiency has gained significant attention with numerous applications (Kolokotsa *et al.* 2011). Simulation and application of artificial intelligence control techniques (such as fuzzy logic and artificial neural networks), has indicated that they have the potential to make significant energy savings in buildings (Kolokotsa *et al.* 2002; Karatasou *et al.* 2006).

Regarding the University campus buildings, the main challenge in the design of control systems is to find the balance between implementation costs, operation costs, energy consumption, indoor environmental quality, user satisfaction and contribution to sustainable living. Intelligent campuses are those involving environmentally responsive design, taking into account the surroundings and building usage, and enabling the selection of appropriate building services and control systems to further enhance the operational building blocks with a view to the reduction of energy consumption and environmental impact over its lifetime (Kolokotsa *et al.* 2016b).

In addition, sensors, actuators and interfaces are essential components for the successful implementation and real-time operation of a web-based energy management system. The evolution of the specific components has been quite rapid during the last decades leading to the intelligent buildings concept, derived from artificial intelligence and information technology.

Based on the analysis above, there is considerable room for improvement and research potential in energy management for group of buildings in a campus context, where different buildings and outdoor spaces are considered.

The School of Design and Environment has three existing buildings (SDE 1–SDE 3) and one building under construction. The interconnection of the SDE buildings in an integrated web-based energy and indoor environmental quality management system will provide a significant boost in the research potential of SDE by developing a Cluster of Living Labs.

To this end, the aim of the present section is to provide guidelines for the implementation of integrated energy management in SDE buildings.

5.7.1. *The phases of the internet-based energy management system for SDE*

The implementation of the internet-based energy management system is split into the phases depicted in Figure 5.27:

1) changes and preparatory work in the existing building energy management systems;

2) wiring and power supply installation;

3) installation of controllers;

4) installation of sensors/actuators;

5) installation of electricity meters;

6) integration and testing.

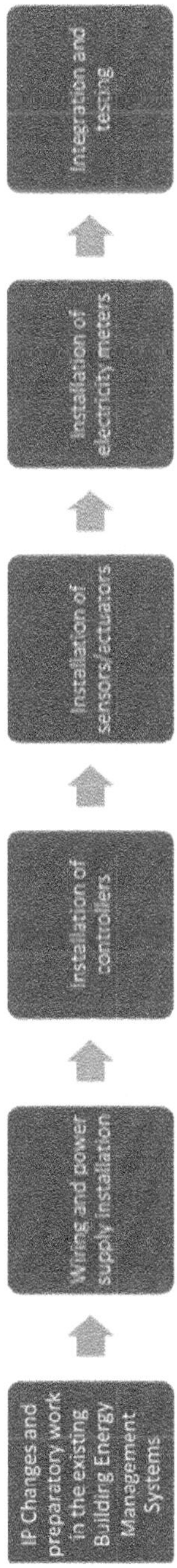

Figure 5.27. *The phases of integrated energy management system implementation*

The changes in the overall hardware architecture is depicted in Figure 5.30. All devices are tabulated in Table 5.6.

The hardware system operation is constantly monitored due to the fact that its characteristics may change over time. The sensors are the main hardware parts whose operational characteristics change over time. The computers, energy meters, HVACs, relays, cabling, etc. have self-diagnostic procedures for faulty operation and are monitored using the software.

The sensor operation drifts over time (1) due to dirt or dust deposited on the sensor, or (2) due to inherent degradation of the sensor behavior. In the case of dirt, regular annual maintenance is scheduled in order to restore the sensor's operation. To cope with degradation of the sensors' functionality, a yearly recalibration of the sensors' characteristics, using reference measurements, is applied. After the recalibration, the appropriate parameters in the control algorithm are adjusted accordingly.

The first phase includes the IP connectivity of HVAC system. An example of such connectivity is provided by the CampIT project[1] for the buildings of the Technical University of Crete. The HVAC is depicted in Figure 5.28. A CS NET web is installed in order to achieve IP/Modbus connectivity of HVAC (see Figure 5.29). The architecture before and after the installation of the internet-based system is depicted in Figure 5.30. The list of possible devices needed is tabulated in Table 5.6.

The communication protocols among the various devices is illustrated in Figure 5.31. The key hardware infrastructure is the automation server and the multi-purpose controller.

The automation server can act as a standalone server and also control I/O Active Directory Domains. It operates over the TCP/IP protocol and supports the Bacnet/Lonworks and Modbus protocols[2].

The second device is the multi-purpose controller. This controller provides:

– ethernet connector for BACnet Ethernet/IP;

– 802.15.4 wireless mesh (25 nodes/network).

It also supports:

– BACnet, EWS, oBIX, FTP;

---

1 Available at: www.campit.gr.

2 Available at: http://www.smarterbuildingtech.com/smartstruxure-universal-input-module-sxwautsvr10001.html.

– EnOcean (wireless);

– ZigBee Pro (wireless);

– Modbus;

– CANbus.

The MPM controller is a device that provides flexibility in the overall connectivity, supporting various communication protocols. Although in the TUC campus the ZigBee Pro is selected, other protocols are also supported by the multi-purpose controller.

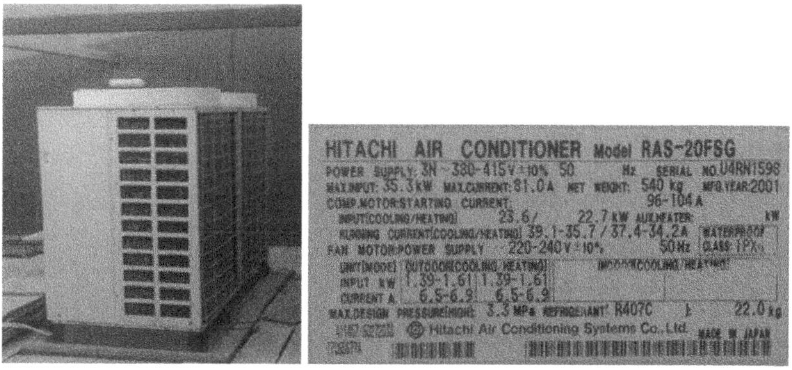

Figure 5.28. *The HVAC system of TUC*

Figure 5.29. *The CS NET WEB*

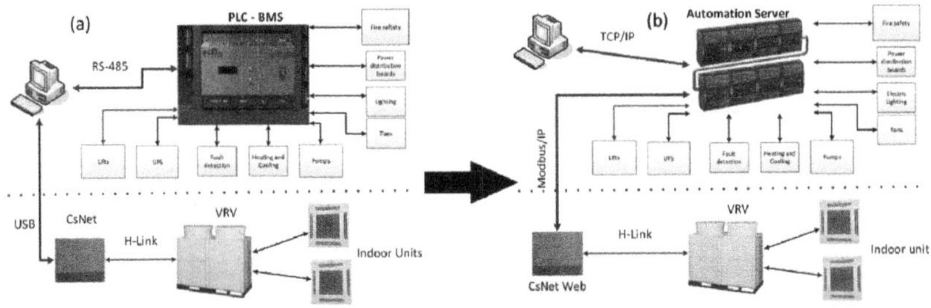

Figure 5.30. *a) The conventional architecture of the energy management, b) the integrated internet-based architecture*

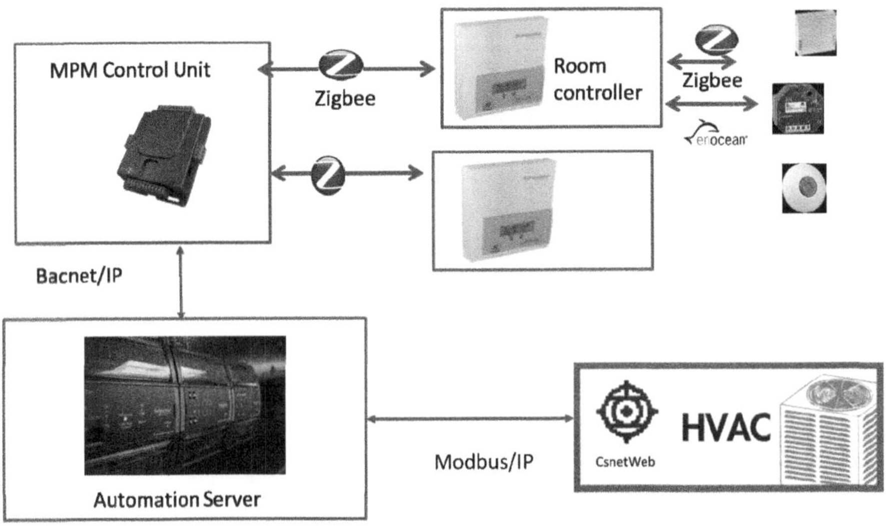

Figure 5.31. *The communication protocols among the various components*

The second phase includes the necessary wiring and power supply and the installation of the multi-purpose controllers, the room controllers and the actuators. The installation of the multi-purpose controller in TUC is depicted in Figure 5.32.

Figure 5.32. *The installation of the multi-purpose controller*

Description	Picture
Multi-Purpose Management device (MPM) enables the control, monitoring and management of entire sites. They can also be used for wired and wireless zone control in larger buildings	
Room controllers (temperature–relative humidity, presence detection sensors)	
CO_2 concentration measurement	
Wireless door switches	

Description	Picture
Wireless window switches	
The sensor SLR 320 converts the illuminance measurement (lux) into 4–20mA or 0–10V	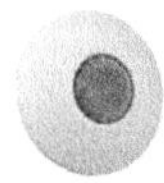
Remote switches	
Presence detector which detects human presence in a range of angles from 0 to 360 degrees against the horizontal plane and with a maximum radius of 7m	
Electricity meter iEM3150	
Electricity meter iEM3155	
Electricity meter iEM3350	
Electricity meter iEM3250 and current transformation	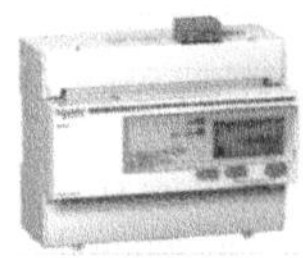
Modbus RTU to Modbus TCP/IP EGX100 Gateway	

Table 5.6. *An indicative list of devices for the integrated energy management. For a color version of this table, see www.iste.co.uk/kolokotsa/smartbuildings2.zip*

5.7.2. *Integration of software and prediction algorithms*

The integration of the various algorithms can be done either using the BCVTB or by programming the multi-purpose controller.

The Building Controls Virtual Test Bed (BCVTB) is a software environment that allows users to couple different simulation programs for co-simulation, and to couple simulation programs with actual hardware. For example, the BCVTB allows the simulation of a building in EnergyPlus and of the HVAC and control system in Modelica, exchanging data between the software as they simulate. The BCVTB is based on the Ptolemy II software environment, and allows expert users of simulation to expand the capabilities of individual programs by linking them to others. Due to the different programs that may be involved in distributed simulation, familiarity with configuring programs is essential. BCVTB can provide links among:

– the EnergyPlus whole building energy simulation program;

– the Modelica modeling and simulation environment Dymola;

– Functional Mock-up Units (FMU) for co-simulation and model-exchange for the Functional Mock-up Interface (FMI) 1.0 and 2.0;

– the MATLAB and Simulink tools for scientific computing;

– the Radiance ray-tracing software for lighting analysis;

– the ESP-r integrated building energy modeling program;

– the TRNSYS system simulation program;

– the BACnet stack, which allows the exchange of data with a BACnet compliant building automation system (BAS);

– the analog/digital interface USB-1208LS from Measurement Computing Corporation (Hodo-Abalo *et al.* 2012) that can be connected to a USB port.

The second option is to program the multi-purpose controller using, for example, the Lua programming environment. Lua is a programming language which can be embedded in different platforms, and is also based on ANSI C.

5.8. Conclusion

The scope of the present work is to reveal the most appropriate energy management techniques for transforming an institutional building to a zero-energy building in the tropics. The various energy management and data-driven techniques already available for transforming a building during its operational phase to a zero-energy building are examined while the role of monitoring is revealed (Kolokotsa *et al.* 2011). Data-driven techniques for building optimisation and control have been used in various building applications (Jain *et al.* 2014; Papantoniou *et al.* 2015). Neural networks are used in the present work as a tool to predict the power loads of buildings in the tropics as well as the outdoor weather conditions, that is, air temperature and relative humidity. The development of an integrated internet-based energy management system for SDE requires the installation of various components to formulate the final architecture. More attention should be paid to the monitoring of relative humidity and latent loads. A possible separation of the sensible and latent loads during the various calculations may help in the development of an integrated energy management system for the tropics. This may be taken into account during the overall refurbishment of the SDE buildings.

6

Objective and Subjective Evaluation of Thermal Comfort in the Loccioni Leaf Lab, Italy

6.1. Introduction

This chapter presents a methodology for the simultaneous subjective and objective evaluation of thermal comfort in commercial buildings. The methodology is utilized for the identification of serious thermal comfort issues, but also for the determination of the preferred comfort conditions in the Loccioni Leaf Lab office spaces, located in the province of Ancona, Italy, during the summer period.

The developed methodology is based on the intermediate protocol level for the evaluation of thermal comfort, as suggested by ASHRAE's Performance Measurement Protocols for commercial buildings (PMP) (Hodo-Abalo *et al.* 2012).

The PMP is intended to provide "a standardized, consistent set of protocols, for a range of cost/accuracy that facilitate the appropriate comparison of the performance of six performance categories (energy, water, thermal comfort, indoor air quality (IAQ), lighting, and acoustics) in commercial buildings, while maintaining acceptable levels of building service for the occupants" (Hodo-Abalo 2012).

Chapter written by Marina LASKARI, Francesco CARDUCCI, Daniela ISIDORI, Martina SENZACQUA, Laura STANDARDI and Cristina CRISTALLI. This chapter is based on a previous work by the authors, "Objective and subjective evaluation of thermal comfort in the Loccioni Leaf Lab", *Energy Procedia*, 134, 645–653, 2017, reproduced with permission from Elsevier.

For a color version of all figures in this chapter, see www.iste.co.uk/kolokotsa/smartbuildings2.zip.

For each of the six performance categories (energy, water, thermal comfort, IAQ, lighting and acoustics) protocols are developed at three levels:

1) *Basic*: basic protocols involve simple, low-cost measures that help provide a preliminary insight into the annual performance of a building and usually involve measurements at the building level.

2) *Intermediate*: intermediate protocols investigate the performance of the building at shorter intervals (i.e. monthly) and at the major system level.

3) *Advanced*: advanced protocols are more detailed and are normally used for the deeper understanding of the building performance (more frequent and more detailed analysis).

This study involves the evaluation of thermal comfort at the *intermediate level.* At the intermediate level, thermal comfort is evaluated using both *objective* (monitoring of indoor comfort parameters) and *subjective* (Right-Now survey) methods.

6.2. Background information

Thermostatically controlled loads in commercial buildings are usually ruled by set points thresholds indicated by international standards, such as the ASHRAE Std. 55 and EN 15251 (ASHRAE 2010b; European Standardisation Organisation 2007). These thresholds set the acceptable comfort range for people across the world, based on several parameters such as the kind of activity, geolocation, external temperature, building orientation, time of the year and so on. However, as it is not easy to capture the actual thermal response of a building, which may change even when compared with similar buildings, it is even harder to standardize the "thermal response" of the people occupying that building. There are so many aspects that can affect the perception of comfort for people, ranging from physical and mental aspects to cultural aspects, and few of them are actually measurable.

Using the approach described in this report, matching objective and subjective measurements, it is possible to draw a cleaner picture of what is the actual, real time comfort level perceived within the building. Information regarding the real time comfort level is necessary to allow for the optimal control of the thermal conditioning system.

Knowing the actual thermal response, of both the building and the people occupying it, is even more important when addressing *demand management strategies*. To correctly evaluate the flexibility given by the thermal loads, it is important to quantify the expected consumption of the building and be aware of the actual acceptable comfort limits. *Demand response strategies* based on *thermostatically controlled loads* use information regarding the internal temperature, external temperature, compressor cycling ratio, and meter data. However, the comfort range is assumed using the aforementioned international guidelines.

Collecting the subjective/objective measurements, enables the training of a *model* to simulate the "*thermal response*" of *people*, therefore obtaining the actual comfort level perceived within the building, based on what people have felt in the past instead of what they should have felt by the guidelines. As a result, a robust model can be produced to *simulate* the thermal response of people to different demand response scenarios, therefore identifying the actual imposable comfort limits.

6.3. Methodology

The methodology developed for this study consists of four steps:

1) collect subjective data from occupants on their thermal comfort;

2) collect concurrent objective measurements of internal and external comfort parameters;

3) make a combined analysis of subjective and objective measurements;

4) determine the Leaf Lab user preferences and satisfaction with internal conditions.

The methodology is outlined in Figure 6.1.

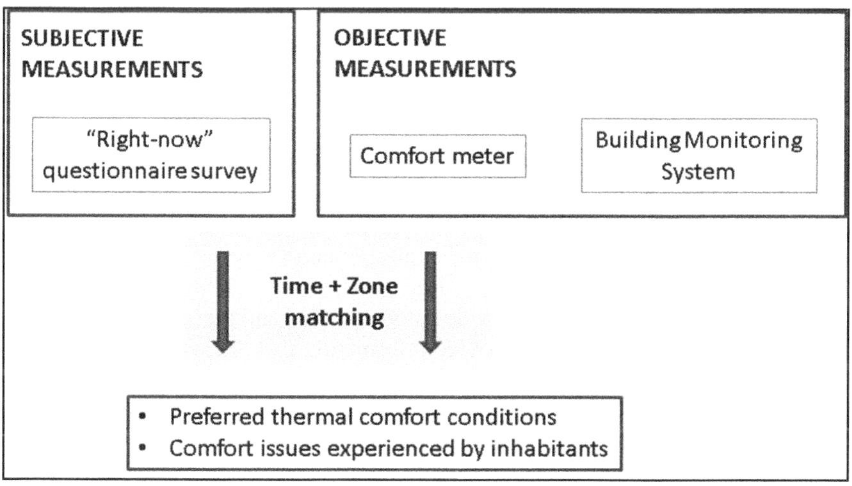

Figure 6.1. *Outline of the study methodology*

6.3.1. *Subjective measurements*

Subjective measurements are collected through an online questionnaire survey. The questionnaire survey is a "Right-Now" survey, in which the date and time is recorded for each response and can therefore be compared against the concurrent monitored indoor environmental conditions.

The survey should be circulated for at least one week so as to capture the possible changes in thermal comfort conditions during the week.

The preferable frequency of response is three times/day:

– morning (on arrival to the office);

– midday (after lunch);

– afternoon (before leaving the office).

At least two people should be assigned responsible for completing the survey each day in each office space, preferably one male and one female each time. Ideally, the same people should respond to the survey every day and at a higher number, however, in this study, we have different volunteers each day.

6.3.2. *Objective measurements*

The objective measurements of indoor environmental conditions are collected in two different ways:

1) standard building monitoring system measurements (MyLeaf system);

2) measurements with a portable comfort meter (Comfort Meter).

The standard building monitoring system measurements give an indication of the conditions under which the HVAC system of the different monitored spaces operates. This is linked to how occupants feel in their thermal environment and therefore to the responses they give in the questionnaire survey.

The portable Comfort Meter provides more detailed and thermal comfort specific measurements and can therefore provide information that is closer to the objectives of this study.

This meter should be installed in the space where most responses to the questionnaire are to be collected. In this study, this space is the Research for Innovation office space (Figure 6.3). The positioning of the Comfort Meter should be made according to the guidelines set in EN 15251:2007 (see Table 6.1).

	Specifications	Explanation
Space	Research for Innovation	– Space in which the seconded people are based and can therefore safeguard the sensor – This is the zone in which more people are likely to take the survey due to interest/relation with the study subject
Height above floor level	~0.6m (can be placed on an office desk)	– Measuring level for sedentary occupants
Location	Center of the room	– Most likely to be away from hot or cold surfaces or direct solar radiation – Representative of room average conditions
Unwanted factors	– Exposure to the sun – Blockage by other objects – Close to warm or hot surfaces (i.e. a computer, a window etc.)	– Would lead to false conclusions for indoor environmental conditions

Table 6.1. *Guidelines for the positioning of the Comfort Meter*

6.3.3. *Combined analysis of objective and subjective measurements*

Through all three measurement processes (building monitoring system; Comfort Meter; questionnaire) the measurement time is available. Therefore, it is possible to match the time that a response to the questionnaire is given with the monitored data. Information on zone characteristics (i.e. floor area, floor level, orientation) that are known to have an impact on the thermal conditions of a space were also collected with the help of a building survey. Furthermore, the zone in which each sensor is located is known, as well as the location of each respondent, thus allowing for zone matching to be performed as well.

The study variables to be considered for the analysis are summarized in Table 6.2. Statistical analysis was performed with the help of Statistical Package for Social Sciences (SPSS) software.

Variable	Variable type*	Building survey	Building monitoring system	Comfort Meter	Questionnaire
Time of day	scale		X	X	x
Period of day	nominal		X	X	x
Gender	nominal				x
Office floor area	nominal	X			
Floor	nominal	X			
Orientation	nominal	X			
Zone (department)	nominal		X	X	x
Outdoor air temperature	scale		X		
Solar irradiance	scale		X		
Indoor air temperature	scale		X	X	
Air velocity	scale			X	
Globe temperature	scale			X	
Radiant temperature	scale			X	
Relative humidity	scale			X	
Predicted mean vote (PMV)	ordinal				x
Predicted mean vote (PMV)	scale			X	
Percentage of people dissatisfied (PPD)	scale			X	
Adjacency to external window	nominal				x
Duration of time sitting at desk	nominal				x
Acceptability	ordinal				x
Thermal preference	ordinal				x

*Nominal: Categorical variable whose values have no intrinsic ranking

Ordinal: Categorical variable whose values have some intrinsic ranking

Scale: Interval or ratio variables. Values represent ordered categories with a meaningful metric.

Table 6.2. *Study variables and source of information*

6.3.4. *User preferences and satisfaction with internal conditions*

Multivariate analysis will be applied on the combined objective and subjective measurements in order to define user preferences regarding internal environmental conditions, and whether any major thermal dissatisfaction issue occurs within the office spaces.

Such findings can prove to be a valuable asset for the facilities or energy manager of the building, as they can provide insights for the optimization of the HVAC system, which in turn can result in the identification of energy saving opportunities and in the improvement of the thermal comfort, and therefore of the productivity of the occupants.

6.4. Collection of building background data

The building studied under this subject is the Loccioni Leaf Lab. It is located in a rural area of the Ancona province in Italy (see Figure 6.2).

Figure 6.2. *Location of the Loccioni Leaf Lab*

The Loccioni Leaf Lab is a two storey building consisting of many different spaces, mainly: laboratories, warehouses, office spaces and meeting rooms. For this study, the following four main office spaces were considered:

1) Research for Innovation (RforI);

2) Energy;

3) Industry;

4) Progettazione.

These office spaces are positioned perimetrically in the building. The Research for Innovation and Energy spaces have a south-east orientation, while the Industry and Progettazione spaces have a north-west orientation (see Figure 6.3).

The characteristics of the four office spaces were collected through a building survey and are summarized in Table 6.3.

We note that, for the heating/cooling operating conditions, although direct user override is not possible, the users can make requests to the facilities management for an increase or decrease of temperature in their specific work area.

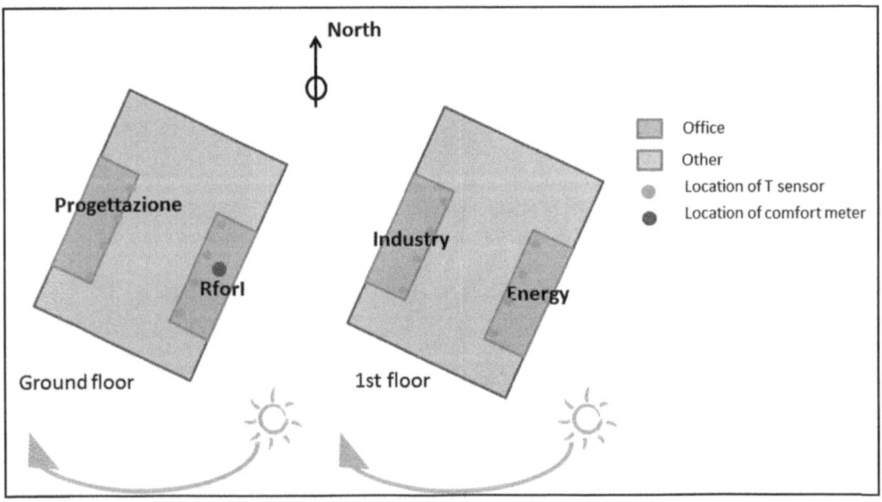

Figure 6.3. *Office spaces considered for the study and location of monitoring equipment*

			RforI	Energy	Industry	Progettazione
Level/Floor			0	0	1	1
Orientation			SE	NW	SE	NW
Monitoring	**Internal IEQ parameters (MyLeaf)**	T air (°C)	Y	Y	Y	Y
		internal lighting power (kW)	Y	Y	Y	Y
	External environmental parameters	T air (°C)	Y	Y	Y	Y
		solar irradiance (W/m^2)	Y	Y	Y	Y
	Number of internal monitoring equipment	T air (°C)	4 sensors	4 sensors	4 sensors	4 sensors
	Measurements frequency		15 min	15 min	15 min	15 min
Occupancy	**Profiles**	Operating hours	07.00-20:00	07.00-20:00	07.00-20:00	07.00-20:00
		Working hours	08.30-18:00	08.30-18:00	08.30-18:00	08.30-18:00
	Occupancy	Occupancy (# people)	15	20	15	10
HVAC system	**Operation schedule**	Days on (from day to day)	Monday–Friday	Monday–Friday	Monday–Friday	Monday–Friday
		Days off (weekends, holidays etc.)	weekends, holidays	weekends, holidays	weekends, holidays	weekends, holidays
		preconditioning (Y/N)	Y	Y	Y	Y
	Location of thermostat	T air (°C)	See Figure 6.3	See Figure 6.3	See Figure 6.3	See Figure 6.3
	Heating operating conditions	heating setpoint (°C)	22	22	22	22
	Cooling operating conditions	cooling setpoint (°C)	26	26	26	26
	Special conditioning requirements	Y/N	N	N	N	N

<table>
<tr><th></th><th></th><th></th><th>RforI</th><th>Energy</th><th>Industry</th><th>Progettazione</th></tr>
<tr><td rowspan="10">Controls</td><td rowspan="4">Fixed setting (user cannot interfere)</td><td>heating (Y/N)</td><td>Y</td><td>Y</td><td>Y</td><td>Y</td></tr>
<tr><td>cooling (Y/N)</td><td>Y</td><td>Y</td><td>Y</td><td>Y</td></tr>
<tr><td>ventilation (Y/N)</td><td>Y</td><td>Y</td><td>Y</td><td>Y</td></tr>
<tr><td>lighting (Y/N)</td><td>Y</td><td>Y</td><td>Y</td><td>Y</td></tr>
<tr><td rowspan="4">Possibility for user override of controls</td><td>heating (Y/N)</td><td>N</td><td>N</td><td>N</td><td>N</td></tr>
<tr><td>cooling (Y/N)</td><td>N</td><td>N</td><td>N</td><td>N</td></tr>
<tr><td>ventilation (Y/N)</td><td>N</td><td>N</td><td>N</td><td>N</td></tr>
<tr><td>lighting (Y/N)</td><td>N</td><td>N</td><td>N</td><td>N</td></tr>
<tr><td rowspan="2">Alternative options for user to adapt the environment</td><td>external operable windows</td><td>Y</td><td>Y</td><td>Y</td><td>Y</td></tr>
<tr><td>external operable doors</td><td>N</td><td>N</td><td>N</td><td>N</td></tr>
</table>

Table 6.3. *Characteristics of the Loccioni Leaf Lab office spaces*

6.5. Collection of monitored data

The thermal environment monitoring was conducted using 2 different methods:

1) standard building monitoring system measurements (MyLeaf);

2) measurements with a portable comfort meter (Comfort Meter).

The *MyLeaf* monitoring system is a monitoring system developed by the Loccioni Group and implemented in its own buildings. Monitoring data for all kinds of parameters are monitored and are accessible through the MyLeaf web-based platform (Figure 6.4). The monitored environmental parameters useful for this analysis are *Indoor Air Temperature* (°C) and *Outdoor Air Temperature* (°C). Four Indoor Air Temperature sensors are located in each office space (Figure 6.3); the measurement frequency is 15 minutes and all measurements can be downloaded from the MyLeaf platform (Figure 6.4) in tabulated form (.csv).

The Comfort Meter is a portable monitoring device designed and developed by the Loccioni Group and the measurements provided are:

– air velocity (m/s);

– air temperature (°C);

– globe temperature (°C);

– radiant temperature (°C);

– relative Humidity (%).

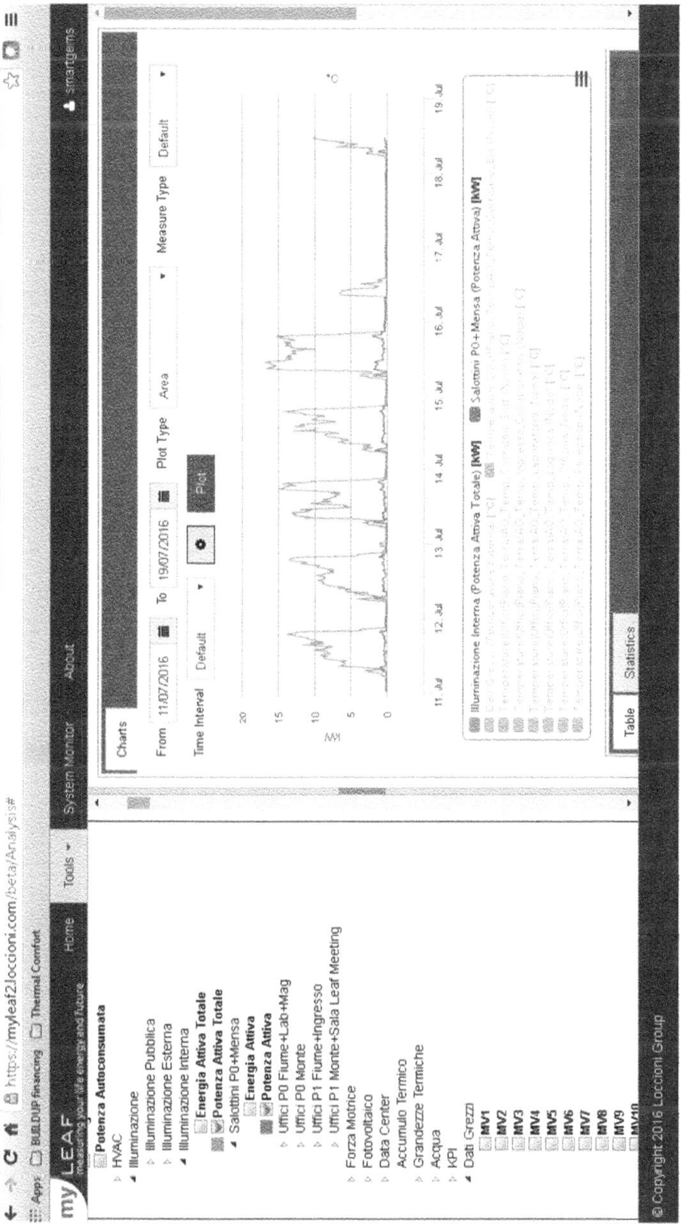

Figure 6.4. Screenshot of MyLeaf platform

After inputting the clo-value (clothing insulation level) and metabolic rate (activity level) of the office space users, the Comfort Meter is capable of calculating the corresponding:

– predicted mean vote (PMV);

– percentage of people dissatisfied (PPD).

A near sedentary activity was assumed (1.2 met) and light office wear (0.5 clo) for the Loccioni LeafLab office spaces.

The sampling time is 10 seconds. The Comfort Meter is connected to a PC and monitored data is saved on the PC in the form of an excel file for each monitored day.

6.6. Right-Now questionnaire survey

There are *two types* of *Right-Now surveys*: self-initiated and requested. For both surveys, the date and time is recorded and can therefore be compared against the concurrent monitored indoor environmental conditions;

– In the self-initiated survey, the occupant can respond to the survey at any given moment through a continuously available link to the survey, without being reminded by a managing department.

– In the requested surveys, the initiative for responding to the survey is taken by a department within the building/company. This is usually in the form of a prescribed reminder email.

Because of the busy schedule of the respondents, the requested survey was considered more appropriate.

The survey was created online using Google forms. The survey was circulated from Friday (22/07) until Friday (29/07) to help monitor changes between Friday and Monday, during which the office spaces are unoccupied and the HVAC system is off. The targeted frequency of response was three times a day:

– morning (arrival at office – 12:30);

– midday (12:30 – 16:00);

– afternoon (16:00 – end of shift).

At least two assigned people in each zone were responsible for completing the survey each day. In order to design a survey that provides only the necessary information regarding thermal comfort, the following questions were selected:

1) Zone: the office and laboratory zones are studied: Research for Innovation, Energy, Industry, and Progettazione.

2) Adjacency to external window (within 3 meters from the respondents' desk).

3) Duration of time sitting at desk (<15 minutes, 5–15 minutes, >15 minutes).

4) Thermal sensation: numerical vote on the ASHRAE seven-point thermal sensation scale (cold, cool, slightly cool, neutral, slightly warm, warm, hot).

5) Acceptability: can be asked in a binary or continuous way. It involves the thermal sensation reported in the previous question. The continuous scale (Clearly acceptable/Just acceptable/Just unacceptable/Clearly unacceptable) allows for more meaningful statistical analysis.

6) Preference: statement of preference for the thermal environment to change or remain the same. Measured on a three-point scale (1-warmer, 2-no change, 3-cooler).

7) Gender: male/female.

The survey does not ask for humidity and radiant effects separately since they are perceived as temperature effects (ASHRAE 2010a). Also, a question on whether thermal comfort enhances the ability of the respondents to get their work done was not included because the building is not naturally ventilated. For this study it is considered that the impact of thermal comfort on productivity is somehow linked to the level of acceptability/unacceptability.

It is worth noting that although activity and clothing level are known to affect thermal sensation, these were not addressed in the survey itself. The reason for not addressing them in the survey is because their values are considered to be rather constant and removing them would help limit the number of questions. A fixed value, relating to the performed task (near sedentary; 1.2 met) and dress code (light office wear; 0.5 clo) was considered for the activity and clothing level. The age range of the Loccioni staff is not very broad (average age is 33–34 years), therefore, a specific question was not included either.

The Right-Now survey was tested before it was formally launched. Only the three questions dedicated to thermal comfort were included in the survey with the purpose of validating their value and suitability for this research as follows (Figure 6.5):

1) Thermal sensation (7-point scale).

2) Acceptability (binary).

3) Preference (3-point scale).

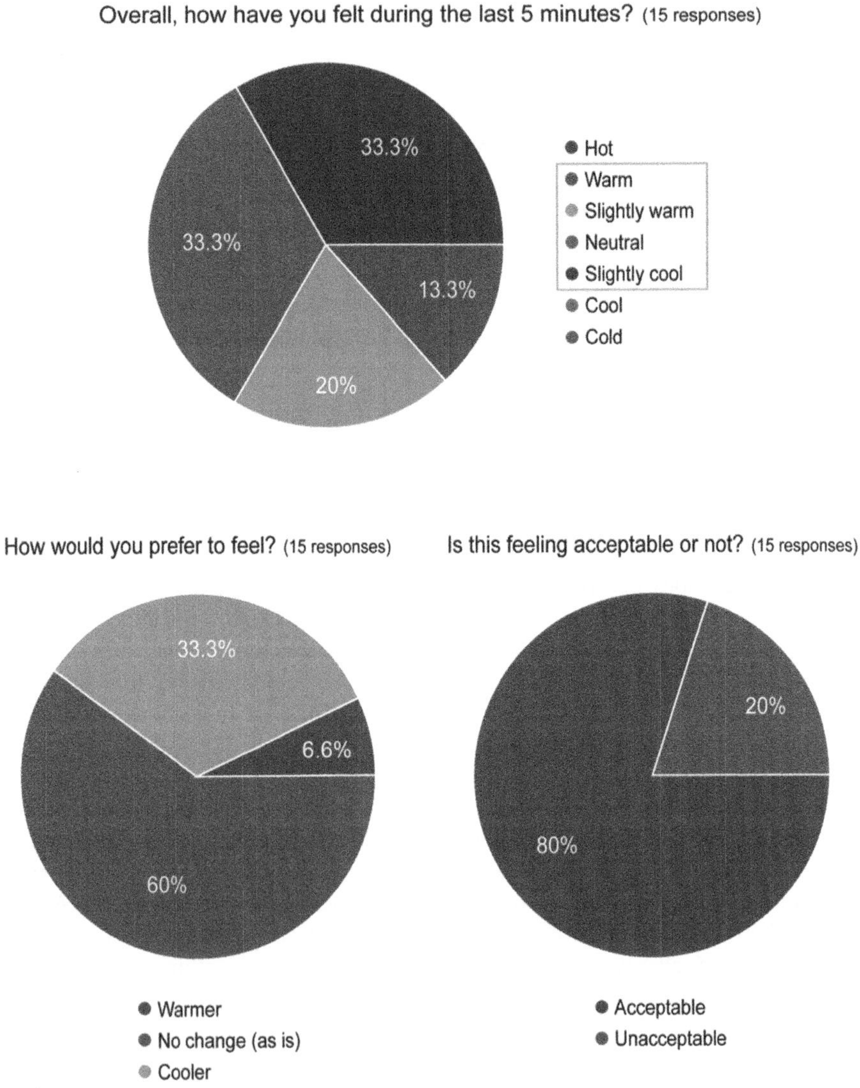

Figure 6.5. *Testing of the Right-Now survey (general statistics)*

The survey was tested in one of the zones for two different, successive days and for the three periods of time during the day that were expected to present differences in thermal sensation (morning, midday, afternoon), in order to show whether it would have meaning to have the survey answered more than once per day (Figure 6.6). Five respondents participated in the test. The respondents were reminded via email to take the survey at all three measurement periods (June 28 afternoon, June 29 morning, June 29 after lunch). The findings of the testing of the "Right-Now" survey are summarised below:

– It is interesting to monitor different times of the day as different conditions (thermal mass behaviour at different times of the day, external conditions, respondent metabolic rate etc.) → the survey should be answered three times/day: morning, midday, afternoon.

– The requested survey (reminders to all via email every day) would soon cause fatigue to the respondents → The survey will be open to everyone to take at any time, but at least two people interested in this activity should be assigned to take the survey every day. Rotational allocation of the task (on a daily basis) would help ensure that the person responding does not get tired of answering the survey three times/day and that a variety of respondents (gender, location in room, age etc.) take the survey.

– The binary option for acceptability (acceptable/unacceptable) does not provide much insight on the level of acceptability/unacceptability. The continuous scale will be used instead.

– Gender should be included as it would be interesting to see the differences in the perception of the thermal environment by different genders.

1	Timestamp	Overall, how have you felt	Is this feeling acceptable	How would you prefer to fe
2	6/28/2016 17:12:34	Slightly cool	Acceptable	No change (as is)
3	6/28/2016 17:19:52	Slightly cool	Acceptable	No change (as is)
4	6/28/2016 17:20:34	Slightly cool	Acceptable	No change (as is)
5	6/28/2016 17:54:00	Neutral	Acceptable	Cooler
6	6/28/2016 18:24:29	Neutral	Acceptable	No change (as is)
7	6/29/2016 10:43:23	Neutral	Acceptable	No change (as is)
8	6/29/2016 10:55:22	Slightly cool	Acceptable	No change (as is)
9	6/29/2016 10:55:28	Neutral	Acceptable	No change (as is)
10	6/29/2016 11:19:49	Slightly cool	Acceptable	Warmer
11	6/29/2016 11:24:21	Neutral	Acceptable	No change (as is)
12	6/29/2016 16:00:32	Warm	Unacceptable	Cooler
13	6/29/2016 16:01:29	Slightly warm	Unacceptable	Cooler
14	6/29/2016 16:05:00	Warm	Unacceptable	Cooler
15	6/29/2016 16:28:09	Slightly warm	Acceptable	Cooler
16	6/29/2016 16:30:38	Slightly warm	Acceptable	No change (as is)

Figure 6.6. *Testing of the Right-Now survey (individual responses)*

6.7. Results

The study of thermal comfort in the Leaf Lab office spaces had a duration of eight calendar days (July 22–29, 2016). The Right-Now survey was circulated in the six working days of the monitoring period. Overall, 214 survey responses were collected in this time. A higher number of respondents were recruited from the Research for Innovation (RforI) office space where the Comfort Meter is located.

Respondent characteristics

The Right-Now survey consists of seven multiple choice questions. Four of them aimed to identify the main characteristics of the respondents which could affect perceived thermal comfort. Those characteristics are summarized in Table 6.4.

A total of 214 responses were collected for the Right-Now survey: 51% of the respondents were from the RforI department, while 15% were from the Energy department and 18% were from the Industry department. In Progettazione, the proportion of respondents was 16%.

There was a good mix of female and male respondents found in the RforI department (54% male, 46% female). In the Energy department, there were 18% female respondents, while in the Industry and Progettazione departments the proportion of female respondents was below 10%. It is worth mentioning that during recruitment of respondents, a good balance between male and female respondents was attempted, however, because the majority of staff in the studied Loccioni departments were male, it was not possible to keep this balance. Nonetheless, the studied male/female ratio can be considered representative of the ratio occupying the space and for which the space is conditioned.

On adjacency to an external window, the majority of respondents (over two-thirds of respondents in all spaces) believed that they were seated within 3 meters from an external window. There are strong indications however that this question was not clearly comprehended by the respondents. The internal wall running the length of all spaces is a glazed wall and it is believed that many of the respondents sitting adjacent to the specific internal wall answered "Yes" to this question.

The time for which the respondents were relaxed in order to adapt to the studied activity (near sedentary) is important in how they perceive the ambient thermal conditions. The majority of respondents in all spaces were seated at their desk for at least 15 minutes before taking the survey. The lowest proportion of respondents seated for at least 15 minutes is found in Research for Innovation (72% of respondents) and the highest in Progettazione (91% of respondents). The highest proportion of respondents seated for less than 15 minutes before taking the survey is found in Energy (18% of respondents). In Progettazione no respondent was seated for less than 15 minutes before taking the survey.

		RforI	**Energy**	**Industry**	**Progettazione**		**Total**
Responses							
	Count	109	33	38	34		214
Gender							
	Male	54%	82%	92%	91%		71%
	Female	46%	18%	8%	9%		29%
Within 3 meters from external window							
	Yes	76%	79%	76%	68%		75%
	No	24%	21%	24%	32%		25%
Time sitting at desk							
	<15 min	9%	18%	8%	0%		9%
	5-15 min	19%	6%	11%	9%		14%
	>=15 min	72%	76%	81%	91%		77%

Table 6.4. *Characteristics of study participants*

6.7.1. *Analysis of MyLeaf measurements*

The MyLeaf system takes measurements on an ongoing basis. Therefore, data is available for the entire duration of the monitoring period at 15 minute intervals. The temperature measurements taken from the MyLeaf system have been analyzed in this section for both the entire duration of monitoring and for the instances that questionnaire responses were collected.

6.7.1.1. *Continuous measurements*

The external and internal air temperatures of the Leaf Lab are monitored at a constant basis through the MyLeaf system. The measurement interval is 15 minutes. Figure 6.7 illustrates the external temperature and the average internal temperature variation in each studied zone for the monitoring period. Although the office spaces are not occupied during weekends, the data for the monitored weekend are considered in the plot to help identify (1) possible changes in internal conditions when the HVAC system is inoperative and (2) whether any major weather changes occurred during the weekend.

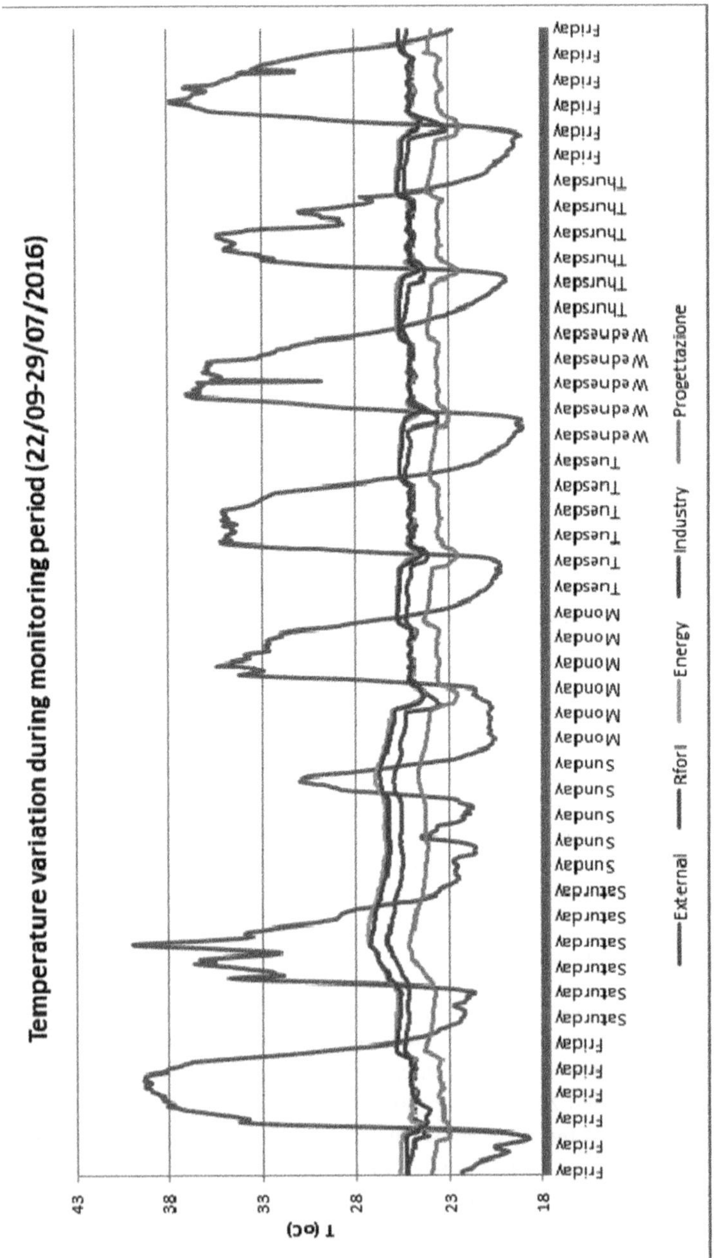

Figure 6.7. *Internal and external temperature diurnal variation between 22/07/2016 and 29/07/2016*

As shown in Figure 6.7, during the hours that the HVAC is inoperative (weekends and weekday non-office hours) the temperature in the office spaces increases. The diurnal variation trend is similar for all office spaces, but the temperature in Progettazione is constantly lower than in the other zones (Table 6.5). This could be attributed to the lower internal gains from lower occupancy, and reduced solar gains due to its west-facing orientation and increased shading from a large tree next to the office space. In any case this is something that should be investigated further in future works. Also, as expected, the temperature in the first floor office spaces (Energy and Industry) is higher than the temperature in the ground floor spaces (Progettazione and RforI). This is more profound during the weekend and during unoccupied hours.

A subject for future research is the preconditioning process of the office spaces that takes place in the early morning hours. A difference is observed in the time that the minimum temperature is reached during the preconditioning process in the ground floor and the first floor zones. The minimum temperature is reached earlier in the ground floor offices. Also, for the RforI space, although the temperature eventually achieved during office hours is similar to that of the first floor spaces, it is still preconditioned to a lower temperature (preconditioning temperature in RforI is lower than the preconditioning temperature in the first floor offices).

A decrease in the external temperature is observed on Sunday. A small impact in the internal temperatures is observed both in the plot in Figure 6.7 and in the mean values summarized in Table 6.5. Namely, the mean values for all of our spaces are lower on Sunday than on Saturday when the external temperature was higher. This is indicative to the reaction that the building has to external conditions when unconditioned.

Day		**RforI**	**Energy**	**Industry**	**Progettazione**	**External**
Friday (22/07)	Mean	24.934	25.004	24.791	23.448	37.001
	Std. Dev.	0.1380	0.1478	0.3878	0.1230	2.3754
	Min	24.5	24.4	24.1	23.1	29.1
	Max	25.2	25.5	25.5	23.8	39.4
Saturday (23/07)	Mean	25.994	26.998	26.861	24.821	33.435
	Std. Dev.	0.3184	0.4009	0.3630	0.3238	2.9212
	Min	25.2	26.1	26.0	23.9	28.5
	Max	26.3	27.4	27.3	25.2	40.0
Sunday (24/07)	Mean	25.724	26.629	26.441	24.429	25.272
	Std. Dev.	0.1820	0.2189	0.1843	0.1767	3.2150
	Min	25.5	26.4	26.3	24.3	21.8
	Max	26.0	27.0	26.8	24.7	31.0

Monday (25/07)	Mean	24.883	24.975	25.018	23.494	30.584
	Std. Dev.	0.2432	0.2064	0.2205	0.3052	4.4392
	Min	24.0	24.4	24.4	22.6	21.6
	Max	25.1	25.4	25.4	23.8	35.5
Tuesday (26/07)	Mean	24.919	25.033	25.032	23.473	33.127
	Std. Dev.	0.1990	0.1130	0.1761	0.2394	2.3388
	Min	24.2	24.8	24.4	22.6	26.6
	Max	25.2	25.4	25.3	23.8	35.2
Wednesday (27/07)	Mean	24.908	25.085	25.036	23.557	33.659
	Std. Dev.	0.2213	0.1114	0.1622	0.1120	2.8719
	Min	24.1	24.8	24.4	23.1	25.8
	Max	25.2	25.4	25.4	23.8	37.0
Thursday (28/07)	Mean	24.927	25.080	25.057	23.477	31.533
	Std. Dev.	0.1599	0.1170	0.1210	0.1809	2.7937
	Min	24.4	24.9	24.8	22.7	27.1
	Max	25.1	25.4	25.4	23.7	35.4
Friday (29/07)	Mean	24.861	25.011	25.011	23.425	34.043
	Std. Dev.	0.3342	0.1504	0.1775	0.2817	2.7790
	Min	23.5	24.6	24.5	22.5	26.3
	Max	25.1	25.3	25.2	23.7	37.9

Table 6.5. *Minimum, maximum, mean values and standard deviations for internal and external temperatures during office hours (22/07/16–29/07/16)*

6.7.1.2. *Matched measurements*

The measurement readings of the MyLeaf sensors were matched to the Right-Now survey responses in terms of time of measurement and zone. The positioning of the sensors and their corresponding IDs are indicated in Figure 6.8. It should be noted that one of the four sensors of the RforI office space (not visible in Figure 6.8) is in an area where no occupants are normally seated and has therefore been excluded from this study.

Figure 6.9 shows the difference between the maximum and minimum temperature in each zone for the instances that the Right-Now survey was answered. The maximum temperature difference in the Progettazione space is close to 2°C. In the Energy department the temperature differences are closer to, but not higher than 1°C. In the RforI and Industry zones the temperature differences are smaller than in the other two zones. Finally, in RforI, a large range in temperature differences compared to the other zones was found.

Figure 6.8. *Location of the MyLeaf sensors in the studied spaces*

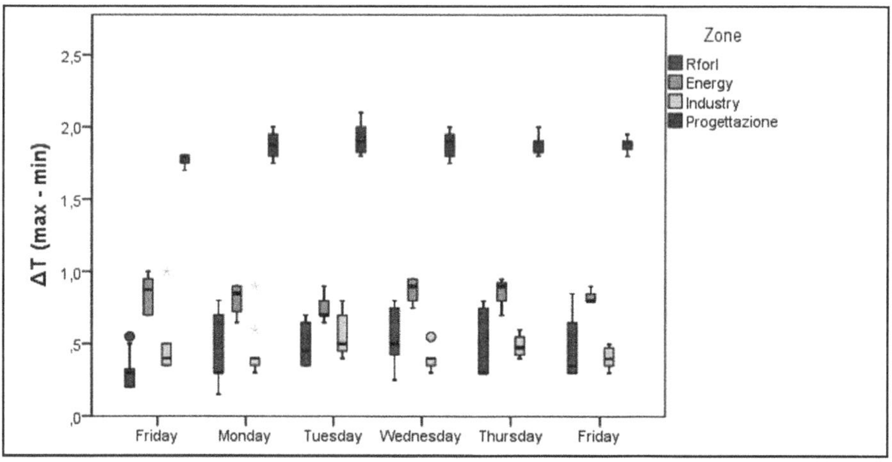

Figure 6.9. *Difference between maximum and minimum sensor reading for matched responses*

As can be determined from Table 6.6, the main issue for the Progettazione space derives from one specific sensor (sensor ID 35a). Nonetheless, the temperatures measured from all four sensors are lower than in any other space. This could be explained by the south-west orientation of the department and therefore the limited solar availability for most of the office hours, and by the blocking of solar penetration from a large tree located in front of the office windows.

Zone_SensorID	Min	Max	Mean	Std. Dev.
Energy_56a	24.40	25.30	25.0363	0.16313
Energy_56b	23.90	24.80	24.6772	0.16866
Energy_57a	24.80	25.70	25.5106	0.15248
Energy_57b	24.50	25.25	25.0576	0.14369
Industry_68ab	24.15	25.10	24.8658	0.24581
Industry_68c	24.50	25.40	25.2262	0.23441
Industry_68d	24.20	25.30	24.9710	0.24067
Industry_68e	24.30	25.30	24.9855	0.20067
RforI_22b	24.20	25.00	24.7220	0.19595
RforI_23a	24.55	25.20	24.9564	0.15292
RforI_23b	24.90	25.40	25.1628	0.13442
Proget_32	22.90	23.90	23.4529	0.20960
Proget_33	23.10	23.95	23.7411	0.16578
Proget_35a	21.45	22.60	22.3426	0.24126
Proget_35b	23.50	24.55	24.2103	0.23605

Table 6.6. *Minimum, maximum, mean values and standard deviations of matched sensor readings*

The average room temperature for each office for the matched survey-monitored measurements was also calculated for different times of day. As shown in Figure 6.10, there is a slight tendency for an increase in internal temperature from morning to midday. In the Energy department the temperature is the most stable during the day, while in the Industry and Progettazione departments the biggest temperature increase is observed from morning to midday (0.23 and 0.19°C, respectively).

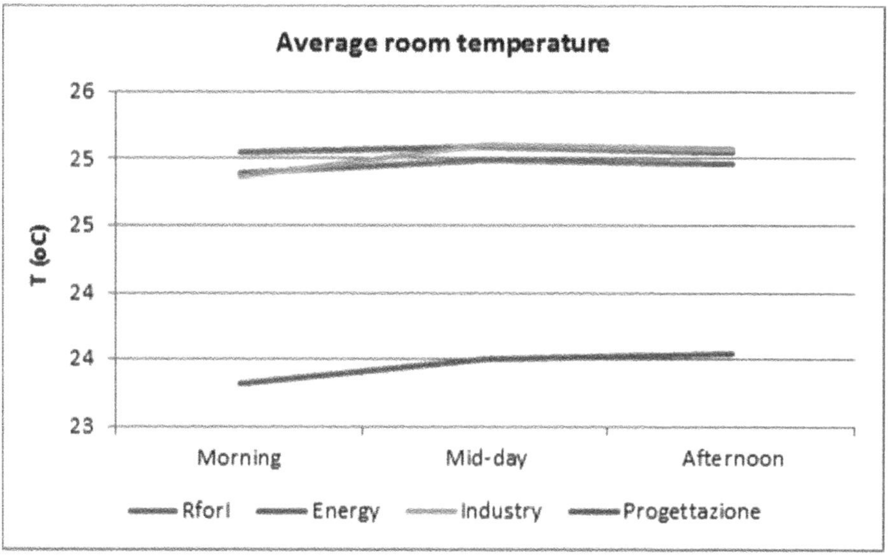

Figure 6.10. *Average room temperature in different office spaces and different times of day for matched responses*

6.7.2. *Analysis of Comfort Meter measurements*

The Comfort Meter took measurements at 10-second intervals for the days that it was installed in the RforI office space. Measurements taken from the MyLeaf system have been analyzed in this section for both the entire duration of monitoring and for the instances that questionnaire responses were collected for the RforI office space.

6.7.2.1. *Continuous measurements*

The Comfort Meter was installed in the RforI office space on the morning of Monday, July 25 and uninstalled on the afternoon of Friday, July 29. Therefore, for Monday, monitored data are available for the midday and afternoon hours, while for Friday, data are available for morning and midday hours. Full day measurements are available for Tuesday and Thursday. On Wednesday, due to connectivity errors between the Comfort Meter and the connected PC, the monitored data were not captured.

A more detailed analysis is performed here for Tuesday and Thursday, for which monitored data are available for all office hours. The data considered are at 10-second intervals. The first measurement is considered as the one taken at 09:18 and the last measurement the one taken at 19:59 on both days.

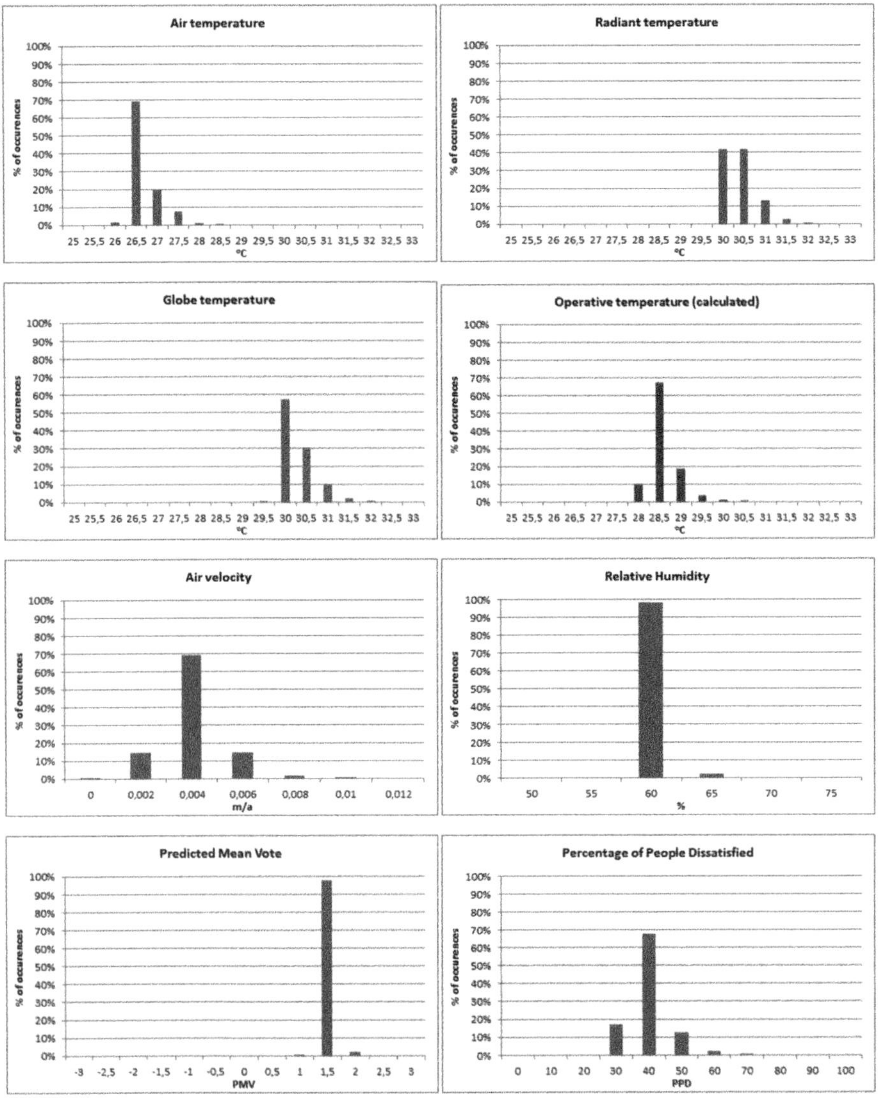

Figure 6.11. *Frequency distribution diagrams for the Comfort Meter measured parameters (Rforl office space)*

Frequency distribution diagrams for all monitored parameters were plotted for Tuesday and Thursday (Figure 6.11) and the results obtained are:

– Relative humidity and air velocity are in line with the criteria ISO 7730 (2006), ASHRAE Std. 55 (2010b) and EN 15251 (2007), with relative humidity below 60% for the majority of the time and air velocity below 0.2 m/s for all instances.

– Radiant temperature and globe temperature are higher than air temperature at all times.

– Because the air speed is lower than 0.2 m/s, the operative temperature was calculated as the average of air temperature and radiant temperature (ASHRAE 2010b). Operative temperature exceeds the EN 15251 (2007) acceptable upper limit of 27 °C at all times.

– PMV is between +1 and +1.5 for the majority of the time, while all PPD values exceed 20%, when the ASHRAE (2010a) criterion for general comfort is -0.5 < PMV < +0.5 and PPD < 10%, respectively.

In summary, the environmental conditions are not compliant with the ASHRAE Standards.

6.7.2.2. *Matched measurements*

In Figure 6.12, the air temperatures occurring at the times that the questionnaire survey was taken (matched objective and subjective measurements) are plotted. As observed, there is a significant offset between air temperatures monitored through MyLeaf and the Comfort Meter. It is worth mentioning that the latter is located on an office desk, while the MyLeaf sensors are placed on metal panels located on glazed internal walls that are about 2 meters away from the closest working station.

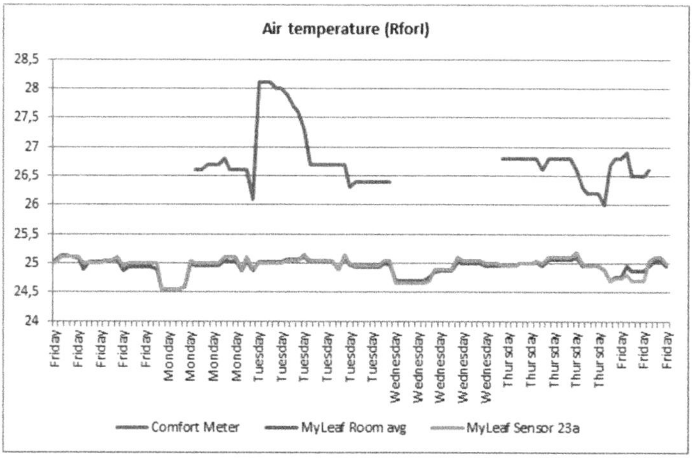

Figure 6.12. *MyLeaf and Comfort Meter measured air temperatures for matched responses (Rforl office space)*

In the RforI office space, air temperature, globe temperature and radiant temperature decrease during the day, with the biggest decrease noted between morning and midday (Table 6.7). Radiant and globe temperature are constantly higher than air temperature (approximately 3.5°C higher on average). Respectively, the PMV improves during the day, but it still remains over +1 (slightly warm). PPD is over 20% at all times. Relative humidity is at its highest in the morning and at its lowest at midday. This could be attributed to the increased presence of people in the morning and the decreased presence of people during the lunch break (midday). Air velocity is at very low levels at all times.

Period of day		Tair (°C)	Tglobe (°C)	Tradiant (°C)	Vair (m/s)	RH (%)	PMV	PPD (%)
Morning	Mean	27.280	30.970	31.010	0.004	60.155	1.452	48.640
	Std. Dev.	0.5745	0.5283	0.5102	0.0015	0.5889	0.1757	9.4055
	Maximum	28.1	31.7	31.9	0.007	61.8	1.7	63.4
	Minimum	26.7	30.4	30.4	0.002	59.6	1.3	39.5
Midday	Mean	26.664	30.268	30.310	0.004	59.404	1.234	37.170
	Std. Dev.	0.1221	0.1406	0.1159	0.0009	0.3780	0.0349	1.7532
	Maximum	26.8	30.5	30.5	0.006	59.9	1.3	40.0
	Minimum	26.3	29.9	30.0	0.002	58.8	1.1	32.1
Afternoon	Mean	26.389	30.058	30.121	0.004	59.474	1.157	33.490
	Std. Dev.	0.1853	0.1774	0.1895	0.0012	0.2446	0.0574	2.7438
	Maximum	26.6	30.3	30.5	0.006	59.8	1.2	37.6
	Minimum	26.0	29.7	29.7	0.002	59.0	1.0	27.7
Total	Mean	26.775	30.425	30.472	0.004	59.659	1.279	39.662
	Std. Dev.	0.4957	0.4954	0.4830	0.0012	0.5389	0.1595	8.3392
	Maximum	28.1	31.7	31.9	0.007	61.8	1.7	63.4
	Minimum	26.0	29.7	29.7	0.002	58.8	1.0	27.7

Table 6.7. *Minimum, maximum, mean values and standard deviations of matched sensor readings (RforI office space)*

6.7.3. *Analysis of Right-Now survey responses*

Descriptive statistics were used to analyze the questionnaire survey data. In the following sections, thermal sensation, acceptability of thermal conditions, and preference for change in thermal conditions are studied for each individual office space and as a whole. Furthermore, analysis is performed to help identify whether respondents with different characteristics perceived the thermal environment in a different way or have different acceptability levels or preferences.

6.7.3.1. *Thermal sensation*

Respondents were asked to select the statement that best describes how they felt during the last five minutes. Options included the typical seven-point ASHRAE thermal sensation scale (-3=cold, 0=neutral, +3=hot).

An overview of total responses to the question is found in Figure 6.13. Half of the total responses suggest a neutral thermal sensation in the office spaces. Another 30% suggest slightly warm conditions, while only 10% suggest slightly cool conditions. Overall, the percentage of votes between -1 < PMV < +1 is 90.7%. Another 7.5% of responses suggest that the thermal sensation at the time that the survey was answered was warm (PMV = 2).

Given that 51% of the respondents are from the RforI, the findings of this figure are driven a great deal by the perception of thermal comfort in the RforI office space. Therefore, the weighted mean based on number of responses from each space, which corresponds to the average thermal sensation of the building, has been calculated and found to be 0.26 ± 0.845. This is in agreement with the ASHRAE PMV range of -0.5 < PMV < +0.5 on acceptable thermal environment for general comfort, which also corresponds to <10 PPD (ASHRAE 2010b).

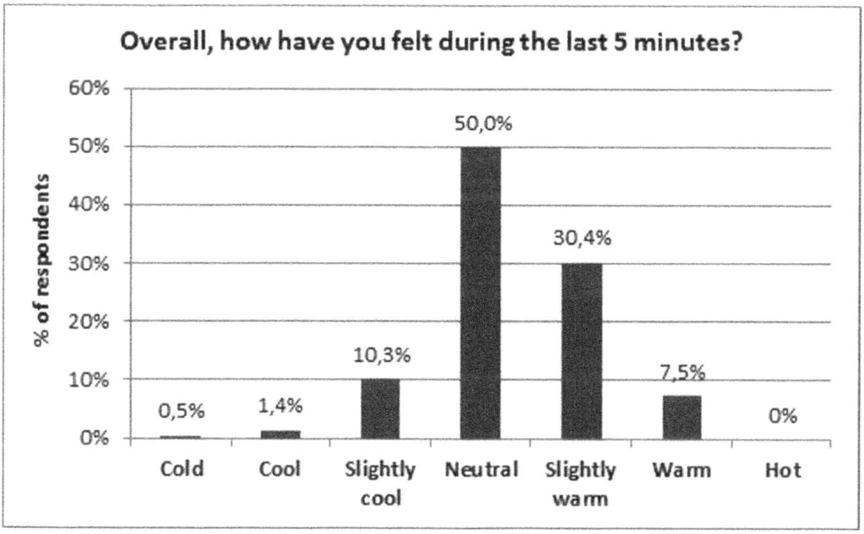

Figure 6.13. *Thermal sensation votes (total sample)*

The biggest proportion of responses in all four spaces suggest a neutral thermal sensation at the time that the survey was answered (Table 6.8). However, a very large number of responses in RforI, Industry and Progettazione suggest a slightly warm thermal sensation (approximately one third of responses in each space). What is interesting to notice is that although the Progettazione department has lower internal temperatures, the thermal sensation of the respondents is similar to that in the other studied spaces. On the contrary, in the Energy department a large proportion of respondents felt slightly cool rather than slightly warm. The Energy department also has the highest proportion of neutral responses. In all individual spaces the mean thermal sensation is between 0.5 and -0.5 (Table 6.9) and is therefore in line with the ASHRAE requirement for <10 PPD.

	Cold	Cool	Slightly cool	Neutral	Slightly warm	Warm	Hot
Rfrol	1%	1%	6%	48%	36%	9%	0%
Energy	0%	0%	27%	64%	3%	6%	0%
Industry	0%	0%	8%	47%	39%	5%	0%
Progettazione	0%	6%	12%	47%	29%	6%	0%

Table 6.8. *Thermal sensation votes (per office space)*

	RforI	Energy	Industry	Progettazione
Mean	0.44	-0.12	0.42	0.18
Std. Dev.	0.84	0.74	0.72	0.94

Table 6.9. *Mean thermal sensation for the monitoring period (per office space)*

In all spaces, except for Industry, the tendency is for respondents to feel cooler in the afternoon than they do in the morning (Figure 6.14). Based on the findings of Figure 6.10, the opposite would be expected; since internal temperature increases slightly during the day, the thermal sensation should change to warmer in the afternoon. Also, it would be expected for the thermal sensation in the Energy department to remain unchanged, since the average room temperature does not vary significantly during the day. In order to investigate this assumption further, measurements such as internal relative humidity and mean radiant temperature are necessary.

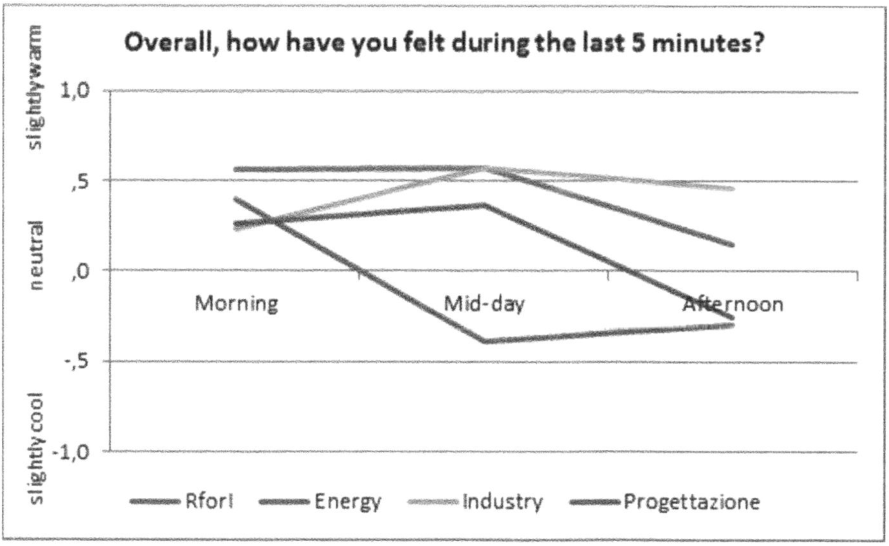

Figure 6.14. *Thermal sensation in different office spaces and different times of day*

6.7.3.2. *Acceptability of current thermal conditions*

Respondents were asked to rate the level of acceptability of the thermal sensation they voted for in the previous question. Options were given on a four-point scale (1= Clearly unacceptable, 2 = Just unacceptable, 3 = Just acceptable, 4 = Clearly acceptable).

An overview of total responses to the question is found in Figure 6.15. Around 91% of responses suggest acceptability with the previously voted thermal sensation. In fact, more than half of the responses suggest clear acceptability. Only 1% of the responses suggest clear unacceptability. The percentage of unacceptability, which can also be translated as percentage of people dissatisfied, is 9% (<10 PPD).

The weighted mean of acceptability, which corresponds to the average acceptability in the building, has been calculated and found to be 1.56 ± 0.637, something between clearly acceptable (+2) and just acceptable (+1).

The biggest proportion of responses in all four spaces suggests acceptability with the thermal environment (Table 6.10). Only in the RforI department is the unacceptability greater than 10% (13% of unacceptability). The reason for unacceptability is probably, based on the findings of Table 6.8, the fact that the thermal environment is slightly warm. In all spaces the mean acceptability is at similar levels (Table 6.11). The highest acceptability is found in Industry

(1.41 ± 0.55) and the lowest in RforI (1.66 ± 0.76), but it should be noted again that the number of responses collected from the RforI are significantly higher (51% of total responses) than those from the other three departments.

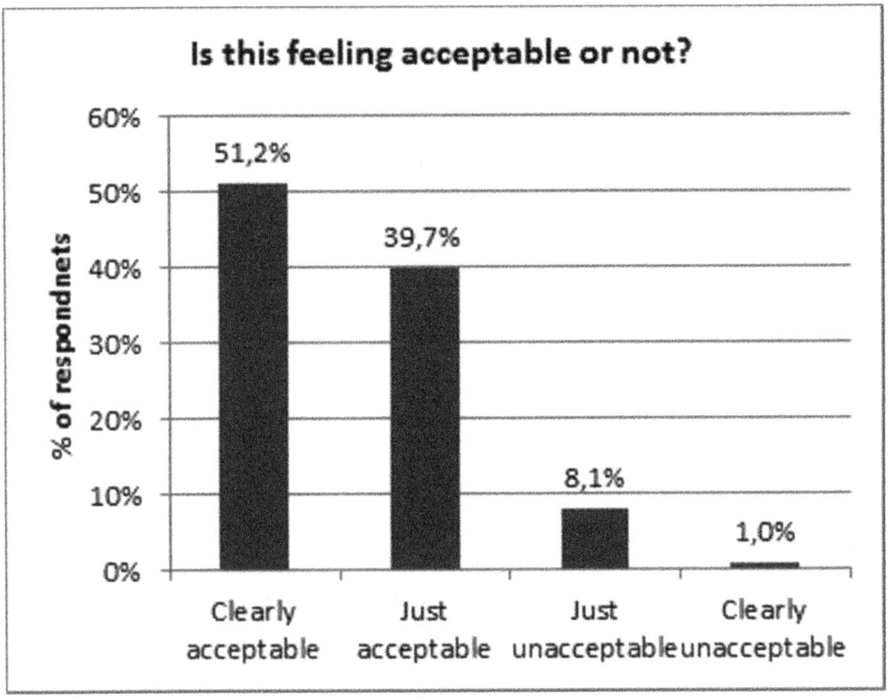

Figure 6.15. *Acceptability of the thermal sensation (total sample)*

	Clearly acceptable	Just acceptable	Just unacceptable	Clearly unacceptable
RfroI	50%	37%	11%	2%
Energy	48%	45%	6%	0%
Industry	62%	35%	3%	0%
Progettazione	47%	47%	6%	0%

Table 6.10. *Acceptability of the thermal sensation (per office space)*

	RforI	Energy	Industry	Progettazione
Mean	1.66	1.58	1.41	1.59
Std. Dev.	0.76	0.61	0.55	0.61

Table 6.11. *Mean acceptability of the thermal sensation (per office space)*

The level of acceptability changes during the day and between office spaces (Figure 6.16). In Progettazione, the level of acceptability decreases as the day progresses, while in Energy it increases. The decrease of acceptability in the Progettazione department is steeper between midday and afternoon. In RforI and Industry, the level of acceptability decreases at midday and increases again in the afternoon. In RforI, however, the increase in the afternoon is steeper.

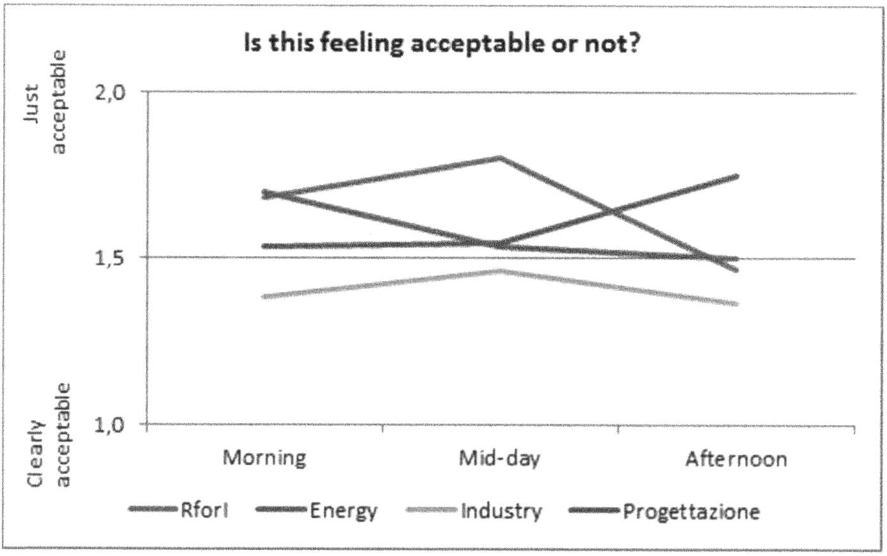

Figure 6.16. *Acceptability of thermal conditions in different office spaces and different times of day*

6.7.3.3. *Preferred thermal sensation*

Respondents were asked to state their preference for change in thermal conditions. Options were given on a three-point scale (-1 = Cooler, 0 = No change, +1 = Warmer).

Overall, 59% of total respondents were satisfied with the thermal sensation they felt when answering the survey, while 36% of them would like to feel cooler (Figure 6.17). Only 4% would like to feel warmer. The weighted mean of preference, which corresponds to the average preference in the building, is found to be -0.30 ± 0.578.

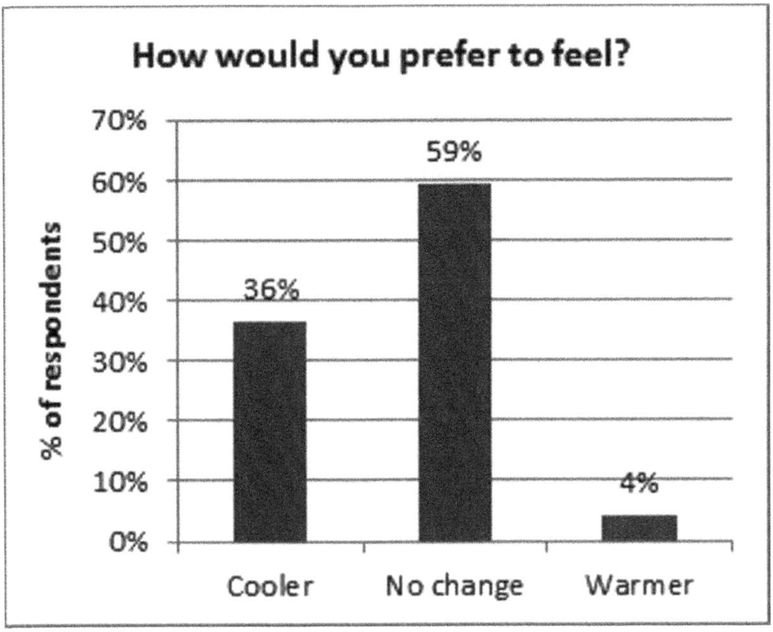

Figure 6.17. *Preference for change in thermal sensation (total sample)*

The biggest proportion of responses in each space is for the thermal environment to remain unchanged. In fact, in the Energy department this proportion is significantly high (88% of responses, Table 6.12) and the mean preference for change is the closest to neutral (Table 6.13). In the remaining three spaces, the proportion of responses stating preference toward a cooler environment is also significant (41–42%) and the mean preferences are toward cooler conditions (negative values). In the Progettazione department, a larger number of responses, compared to the other spaces, is found towards preference for a warmer environment. This is somehow in agreement with findings showing a lower temperature in the Progettazione department, compared to the other departments, at all times.

	Cooler	No change	Warmer
Rfrol	41%	57%	2%
Energy	9%	88%	3%
Industry	42%	53%	5%
Progettazione	41%	47%	12%

Table 6.12. *Preference for change in thermal sensation (per office space)*

	Rforl	Energy	Industry	Progettazione
Mean	-0.39	-0.06	-0.37	-0.29
Std. Dev.	0.53	0.35	0.59	0.68

Table 6.13. *Mean preference for change in thermal sensation (per office space)*

The preference for a warmer environment increases in RforI and Progettazione in the afternoon (Figure 6.18). In Energy, the preference for a warmer environment increases both at midday and in the afternoon. In Industry, there is no significant change in preference during the day.

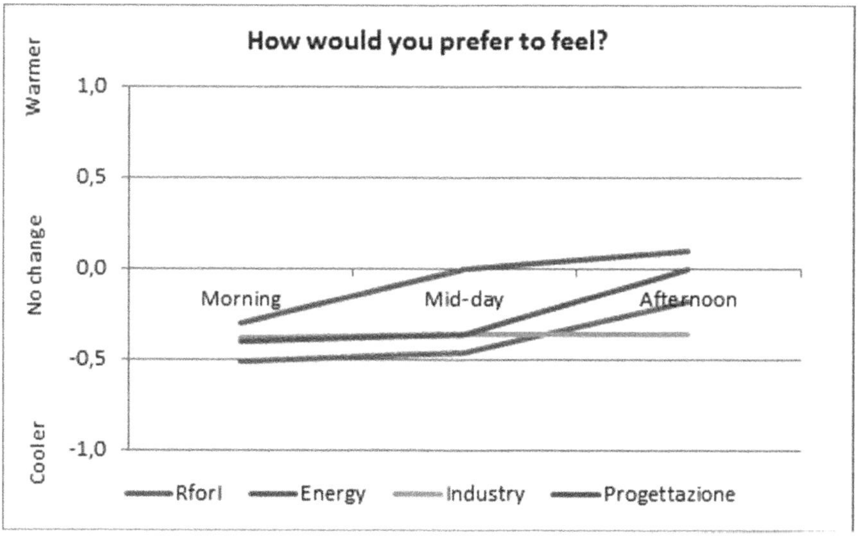

Figure 6.18. *Thermal sensation preference in different office spaces and different times of day*

6.7.4. *Respondent characteristics and thermal comfort*

The responses to the Right-Now survey were further analyzed per respondents' characteristics to determine whether they play a role in how occupants perceive their thermal environment. These characteristics are:

– gender;

– time sitting at desk before taking the survey;

– adjacency to external window.

6.7.4.1. *Gender*

The *thermal sensation* is between -1 < PMV <+1 for 91% of responses coming from female respondents and for 89% from male respondents (Figure 6.19). However, more male respondents feel slightly cooler than neutral, compared to female respondents that are feel more neutral than slightly cool. The proportion of respondents feeling slightly warm is similar for the two genders (31% for male and 29% for female respondents). The mean PMV value for both male and female responses is within the prescribed PMV range of -0.5 < PMV < +0.5 (male 0.33 ± 0.822, female 0.26 ± 0.904).

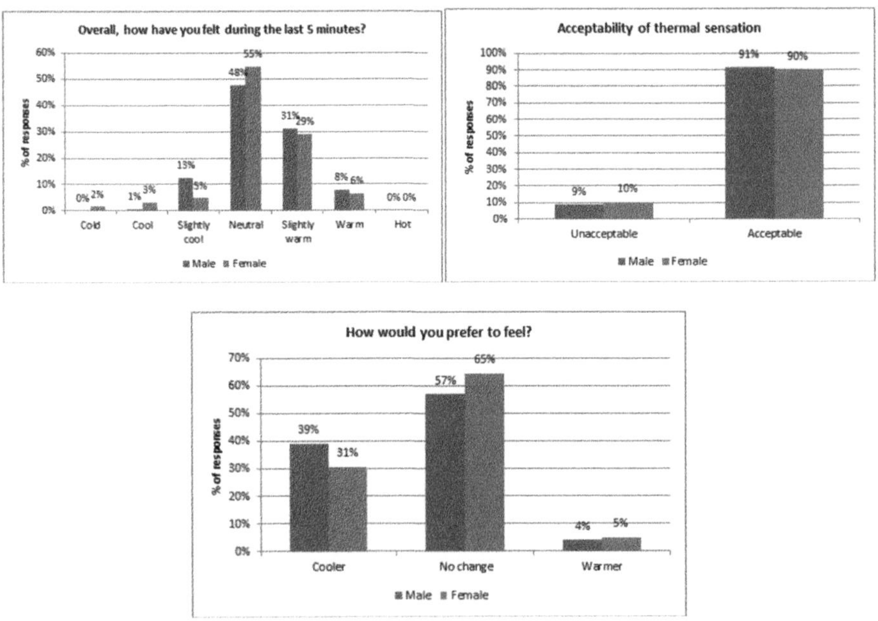

Figure 6.19. *Thermal comfort per gender (total sample)*

There is no significant difference in *acceptability* between genders. However, unacceptability is slightly over 10% for female respondents (10.2%) (>10 PPD).

The majority of both male and female respondents would *prefer* for the thermal conditions to remain unchanged. Although similar proportions of male and female respondents find their thermal sensation acceptable, a higher proportion of male respondents would prefer to feel cooler. Around 39% of male respondents would prefer to feel cooler, while 31% of female respondents would prefer to feel cooler. A much smaller proportion of both male and female respondents would prefer to feel warmer.

6.7.4.2. *Time sitting at desk before taking the survey*

The proportion of respondents with a thermal sensation between -1 < PMV < +1 increases with the time seated at a desk (Figure 6.20). This is 84% for those sitting for less than 5 minutes, 90% for those sitting between 5–15 minutes and 91% for those sitting for 15 minutes or more. However, the proportion of neutral respondents is the lowest for the case of those seated between 5–15 minutes. This group of respondents also has the highest proportion of respondents feeling slightly warm. The mean PMV value for all three groups is near 0.3 and therefore within the prescribed PMV range of -0.5 < PMV < +0.5.

Although it would be expected for those sitting for less time at their desk to find their thermal conditions more unacceptable since they are assumed to have had more intense activity the time before, the findings show that there is no significant difference in the acceptability of thermal sensation between those sitting for less than 5 minutes before taking the survey and those sitting for at least 15 minutes. What is interesting to observe is that those sitting between 5–15 minutes have the highest rate of unacceptability (17% of respondents).

The majority of respondents sitting at their desk for less than 15 minutes and for more than 15 minutes were satisfied with the thermal conditions at the time they took the survey (Figure 6.20). Contrarily, the biggest proportion of respondents sitting at their desk between 5–15 minutes (50% of respondents) would prefer to feel cooler. Only 43% of them were satisfied with conditions as they were. Another 7% of them, the biggest proportion from the three groups, would prefer to feel warmer. No respondent seated for less than 15 minutes would like to feel warmer.

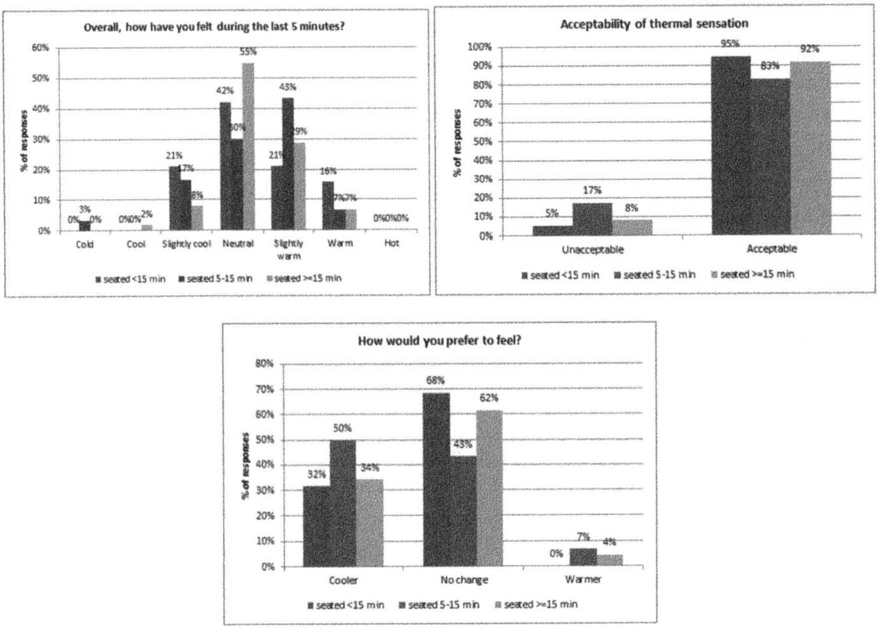

Figure 6.20. *Thermal comfort per time seated (total sample)*

6.7.4.3. *Adjacency to external window*

The thermal sensation is between -1 < PMV < +1 for 91%, for both the respondents sitting within 3 meters from an external window and those not sitting adjacently to an external window (Figure 6.21). Nonetheless, those sitting adjacent to an external window feel warmer rather than cooler. Those not sitting adjacent to a window feel neutral at a higher proportion than those sitting adjacent to a window, and also can feel both warmer or cooler. The mean PMV value for groups of respondents is within the prescribed PMV range of -0.5 < PMV < +0.5 (adjacent to window 0.39 ± 0.824, not adjacent to window 0.08 ± 0.874), but the thermal sensation for those not adjacent to an external window is closer to neutral.

The proportion of respondents sitting within 3 meters from an external window that finds the thermal sensation unacceptable is higher than the proportion of those not sitting adjacently to an external window, and is also slightly over 10% (10.3% of respondents) (>10 PPD).

Both respondents sitting close to an external window and those not sitting close to an external window are satisfied with their thermal environment at a great number (58% and 64% of respondents, respectively). There is a higher proportion of respondents adjacent to an external window that would prefer to feel cooler than respondents not sitting close to an external window (39% and 30% of respondents, respectively). Only a small number of respondents in total would like to feel warmer.

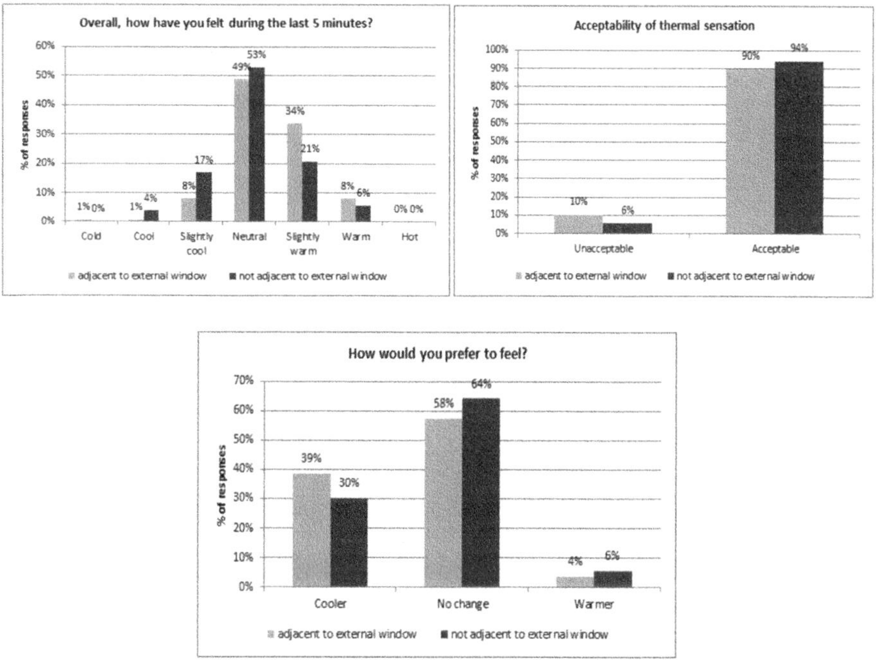

Figure 6.21. *Thermal comfort per desk location (total sample)*

6.7.5. *Combined analysis of objective and subjective measurements*

Multivariate analysis was applied to the combined objective and subjective measurements. The following sections present the findings of this analysis.

6.7.5.1. *Determination of patterns in thermal sensation and average temperature for day of the week and period of day*

Thermal sensation and average room temperature were studied in more detail to help determine any specific patterns for period of day or day of the week.

For the *Research for Innovation* department (Figure 6.22), a cooler thermal sensation is observed for all days in the afternoon hours, compared to the morning and midday hours. The only day when the thermal sensation is similar for all three periods of the day is on Thursday. For the morning and midday hours, there is no specific trend. What is worth noting is the fact that although on Monday and Wednesday morning the average temperature was lower than that for the midday and afternoon hours, the thermal sensation for that time was rather warm.

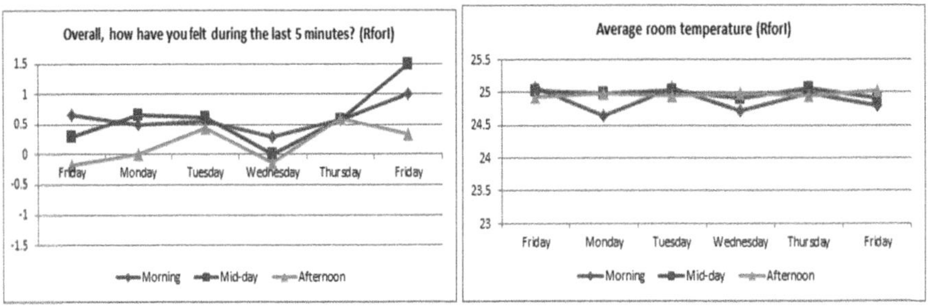

Figure 6.22. *Mean thermal sensation and average internal temperature for different days of the week and different times during the day (Research for Innovation office space)*

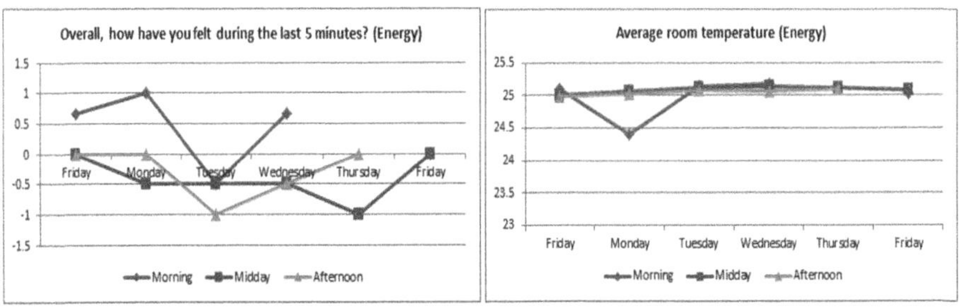

Figure 6.23. *Mean thermal sensation and average internal temperature for different days of the week and different times during the day (Energy office space)*

In the *Energy* department, the thermal sensation in the morning is warmer than in the afternoon from Friday–Wednesday (Figure 6.23). For the last two days of the monitoring period (Thursday and Friday), it is not possible to compare the two due to a lack of responses (note than on Friday the thermal sensation during the morning is the same as for midday). The average temperature is rather similar throughout the day and week. A lower temperature is only observed on Monday morning, which could be related to the preconditioning of the space. Nonetheless the average thermal sensation is the warmest on Monday morning, compared with all of the other days.

It should be noted that both the Energy and the RforI departments have direct sunlight in the morning, thus, even though the building is highly insulated, this may affect the office temperature.

From Friday until Wednesday, the thermal sensation during the midday hours is warmer than in the morning hours in the *Industry* department (Figure 6.24). During the morning hours the thermal sensation is cooler, while in the afternoon the thermal sensation is between or similar to the morning or midday thermal sensation. For the last two days of monitoring no specific trend is observed. The average temperature is rather similar for the midday and afternoon hours. In the morning hours, the average internal temperatures are lower on the first days of monitoring (Friday) and on Monday. After Tuesday it reaches similar values to the midday and afternoon.

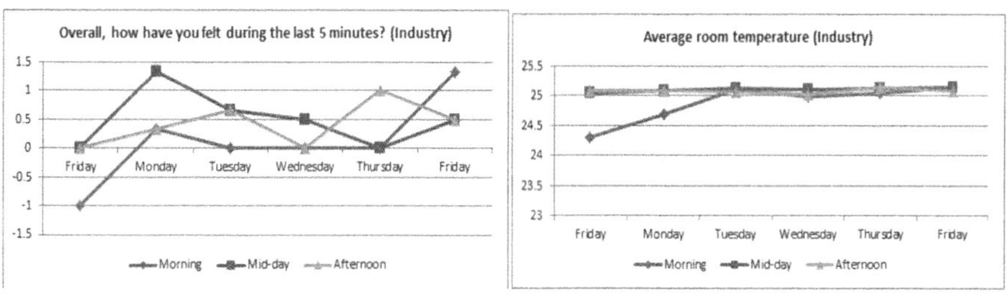

Figure 6.24. *Mean thermal sensation and average internal temperature for different days of the week and different times during the day (Industry office space)*

In the *Progettazione* department a similar trend between midday and morning thermal sensation is observed (Figure 6.25). For the afternoon hours no specific pattern is observed. The average internal temperature is lower and in some cases similar to the midday and afternoon temperatures.

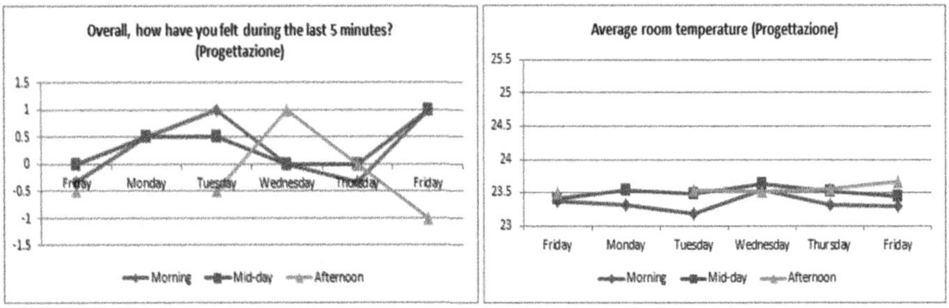

Figure 6.25. *Mean thermal sensation and average internal temperature for different days of the week and different times during the day (Progettazione office space)*

6.7.6. *Correlation analysis for MyLeaf and Right-Now survey measurements*

In the view of determining the existing relation among variables, the objective measurements and the answers from the survey, correlation analysis was performed. Spearman's correlation coefficient *(rho)* was used in this analysis instead of the Pearson correlation, as this is the measure normally used when not all variables are normally distributed and outliers may be present.

The correlation matrix for the analysis performed on the MyLeaf and Right-Now survey variables is found in Table 6.14. The correlations worth reporting are those at the significance levels of 0.01 and 0.05, and are highlighted in the table in green and orange, respectively. In more detail, correlations are found between the following:

– *Thermal sensation* has a moderate positive correlation with acceptability ($r = 0.447$, $p < .001$) and a strong negative correlation with preference ($r = -0.763$, $p < .001$). A weaker correlation, at a lower significance level ($p < .005$), is also found with adjacency to external window, period of day and floor level.

– *Acceptability* of the thermal sensation has a moderate negative correlation with preference for change of the thermal conditions ($r = -0.528$, $p < .001$).

– *Preference* for change of the thermal conditions has a weak positive correlation with period of day ($r = 0.203$, $p < .001$).

– It is worth highlighting that none of the three abovementioned variables related to thermal comfort are found to be correlated to *internal temperature.* Internal temperature (average, minimum and maximum) has a weak correlation, yet a high significance level ($p < .001$), with the zone related characteristics of orientation and floor level. Minimum temperature has a weak positive correlation ($r = 0.274$, $p < .001$) with period of day.

– *External temperature* is not found to be correlated to subjective thermal comfort measurements.

6.7.7. Correlation analysis for objective and subjective measurements (Research for Innovation office space)

Spearman's correlation was also used for the correlation analysis between all objective and subjective measurements. This analysis is only possible for the Research for Innovation office space in which the Comfort Meter was installed. The correlation matrix is found in Table 6.15. Correlations at the significance levels of 0.01 and 0.05 are highlighted in the table in green and orange, respectively. In more detail, correlations are found between the following:

– *Thermal sensation*, *Acceptability* and *Preference* are only correlated between them. However, the correlation between thermal sensation and acceptability and between acceptability and preference ($r = 0.638$, $p < .001$ and $r = -0.667$, $p < .001$, respectively) is stronger than the one found Table 6.14. The correlation strength between thermal sensation and preference ($r = -0.727$, $p < .001$) is similar to that found in Table 6.14. Although the Comfort Meter calculated the PMV, this calculated PMV is not found to be correlated to the thermal sensation voted by the survey respondents.

– The *air temperature* monitored through the Comfort Meter is correlated to the mean average temperature monitored via the MyLeaf system ($r = 0.482$, $p < .001$). A weaker correlation between the air temperature monitored through the Comfort Meter and the air temperature measured by the MyLeaf sensor closest to the Comfort Meter (Sensor 32a) is also found ($r = 0.268$, $p < .005$). The MyLeaf average temperature is correlated to all the Comfort Meter measured parameters (incl. PMV and PPD) except Air Velocity.

– *Period of day* is negatively correlated to all of the Comfort Meter measured parameters (incl. PMV and PPD) except Air Velocity. In fact, the absolute r value is greater than 0.8 for the cases of T air, T globe, T radiant, PMV and PPD.

– *Weekday* is correlated to Relative Humidity ($r = 0.482$, $p < .001$) and to the temperature measured by Sensor 32a ($r = -0.250$, $p < .005$).

– The *external temperature* is only correlated to the average internal temperature resulting from the MyLeaf system measurements ($r = 0.302$, $p < .05$).

Spearman's rho	Thermal sensation	Acceptability	Preference	Office space	Gender	Adjacency to ext. window	Time sitting at desk	Period of day	Weekday	Orientation	Floor	Tavg	Tmax	Tmin	ΔT (Max-Min)	Text	
Thermal sensation	r	1															
	Sig.	.															
Acceptability	r	0.447**	1														
	Sig.	0.000	.														
Preference	r	-0.763**	-0.528**	1													
	Sig.	0.000	0.000	.													
Office space	r	-0.119	-0.063	0.069	1												
	Sig.	0.083	0.363	0.317	.												
Gender	r	-0.016	0.064	0.079	-0.379**	1											
	Sig.	0.821	0.356	0.253	0.000	.											
Adjacency to ext. window	r	-0.167*	-0.076	0.082	0.043	0.108	1										
	Sig.	0.015	0.276	0.235	0.529	0.118	.										
Time sitting at desk	r	-0.015	-0.076	0.063	0.155*	0.016	0.075	1									
	Sig.	0.825	0.278	0.359	0.023	0.816	0.278	.									
Period of day	r	-0.141*	-0.07	0.203**	-0.043	-0.034	0	-0.004	1								
	Sig.	0.039	0.314	0.003	0.534	0.621	1	0.959	.								
Weekday	r	-0.04	0.112	-0.078	0.027	-0.008	-0.014	-0.086	-0.012	1							
	Sig.	0.56	0.107	0.258	0.698	0.91	0.836	0.212	0.859	.							
Orientation	r	0.012	-0.081	-0.034	0.885**	-0.327**	0.048	0.153*	-0.049	0.002	1.000						
	Sig.	0.858	0.246	0.621	0.000	0.000	0.487	0.025	0.478	0.973	.						
Floor	r	-0.148*	-0.089	0.132	0.463**	-0.256**	-0.038	0.009	0.038	-0.02	0.296**	10.000					
	Sig.	0.031	0.200	0.055	0.000	0.000	0.577	0.899	0.580	0.767	0.000	.					
Tavg	r	0.029	0.027	0.034	-0.174*	-0.052	-0.091	-0.068	0.103	-0.001	-0.228**	0.558**	1				
	Sig.	0.677	0.699	0.622	0.011	0.446	0.186	0.326	0.133	0.983	0.001	0.000	.				
Tmax	r	-0.03	0.042	0.074	-0.173*	-0.01	-0.105	-0.056	0.027	0.018	-0.332**	0.614**	0.866**	1			
	Sig.	0.666	0.543	0.28	0.011	0.879	0.128	0.417	0.689	0.793	0.000	0.000	0	.			
Tmin	r	0.112	-0.021	0.003	-0.362**	0.043	-0.08	-0.074	0.274**	0.033	-0.228**	0.241**	0.774**	0.506**	1		
	Sig.	0.103	0.766	0.962	0.000	0.532	0.247	0.281	0	0.631	0.001	0.000	0.000	0.000	.		
ΔT(MAX-MIN)	R	-0.087	0.067	0.031	0.592**	-0.180**	0.01	0.149*	-0.273**	-0.016	0.335**	0.101	-0.313**	-0.039	-0.747**	1	
	Sig.	0.207	0.334	0.647	0.000	0.008	0.887	0.03	0.000	0.816	0.000	0.141	0.000	0.568	0.000	.	
Text	R	-0.073	0.078	-0.002	0.001	-0.005	0.034	-0.251**	0.101	0.414**	-0.013	0.076	0.197**	0.122	0.242**	-0.217**	1
	Sig.	0.290	0.260	0.979	0.988	0.947	0.625	0.000	0.141	0.000	0.850	0.266	0.004	0.074	0.000	0.001	.

**Correlation is significant at the 0.01 level (2-tailed).

*Correlation is significant at the 0.05 level (2-tailed).

Table 6.14. *Correlation matrix for studied variables (MyLeaf measurements and Right-Now survey responses) (N=214). For a color version of this table, see www.iste.co.uk/kolokotsa/smartbuildings2.zip*

Spearman's rho	Thermal sensation	Acceptability	Preference	Gender	Adjacency to ext. window	Time sitting at desk	Period of day	Weekday	CM_Ta	CM_Tg	CM_Tr	CM_Va	CM_RH	CM_PMV	CM_PPD	Tavg, MyLeaf	Sensor 23a	Text	
Thermal sensation	r	1																	
	Sig	.																	
Acceptability	r	0.638**	1																
	Sig	0.000	.																
Preference	r	-0.727**	-0.667**	1															
	Sig	0.000	0.000	.															
Gender	r	-0.124	0.081	0.09	1														
	Sig	0.328	0.527	0.482	.														
Adjacency to ext. window	r	-0.177	-0.095	0.138	0.280*	1													
	Sig	0.165	0.462	0.281	0.026	.													
Time sitting at desk	r	-0.1	-0.096	0.063	0.075	0.02	1												
	Sig	0.433	0.453	0.622	0.554	0.875	.												
Period of day	r	-0.177	-0.136	0.228	0.021	-0.028	-0.043	1											
	Sig	0.163	0.289	0.07	0.871	0.828	0.737	.											
Weekday	r	0.193	0.135	-0.239	0.055	-0.008	0.07	-0.222	1										
	Sig	0.129	0.297	0.059	0.669	0.95	0.585	0.081	.										
CM_Tair	r	0.06	0.06	-0.063	0.026	-0.055	0.156	-0.859**	0.015	1									
	Sig	0.638	0.643	0.622	0.84	0.667	0.219	0.000	0.907	.									
CM_Tglo	r	0.078	0.025	-0.05	-0.002	-0.018	0.182	-0.839**	0.031	0.953**	1								
	Sig	0.539	0.849	0.693	0.989	0.889	0.151	0.000	0.812	0.000	.								

CM_Trad	r	0.128	0.095	-0.093	-0.028	-0.023	0.212	-0.811**	0.109	0.906**	0.941**	1							
	Sig	0.312	0.458	0.465	0.827	0.857	0.093	0.000	0.396	0.000	0.000	.							
CM_Vair	r	-0.064	0.021	0.211	-0.107	0.129	-0.223	-0.002	-0.11	0.005	0.065	0.006	1						
	Sig	0.614	0.869	0.094	0.399	0.314	0.077	0.986	0.392	0.968	0.61	0.964	.						
CM_RH	r	0.172	0.183	-0.238	-0.026	0.015	-0.011	-0.562**	0.482**	0.315*	0.342**	0.415**	0.007	1					
	Sig	0.173	0.151	0.058	0.841	0.908	0.932	0.000	0.000	0.011	0.006	0.001	0.959	.					
CM_PMV	r	0,112	0,087	-0,104	-0,004	-0,037	0,204	-0,874**	0,095	0,973**	0,967**	0,971**	-0.01	0.410**	1				
	Sig	0.377	0.499	0.412	0.973	0.775	0.106	0.000	0.457	0.000	0.000	0.000	0.938	0.001	.				
CM_PPD	r	0.108	0.085	-0.103	-0.008	-0.035	0.199	-0.877**	0.096	0.974**	0.967**	0.969**	-0.008	0.414**	0.999**	1			
	Sig	0.394	0.508	0.416	0.952	0.787	0.116	0.000	0.456	0.000	0.000	0.000	0.947	0.001	0.000	.			
Tavg, MyLeaf	r	-0.116	-0.113	0.167	0.114	-0.145	0.099	-0.167	-0.197	0.482**	0.419**	0.403**	0.098	-0.339**	0.421**	0.416**	1		
	Sig	0.36	0.378	0.187	0.37	0.258	0.438	0.187	0.121	0.000	0.001	0.001	0.44	0.006	0.001	0.001	.		
Sensor 23a	r	-0.123	-0.125	0.162	0.065	-0.143	0.112	0.046	-0.250*	0.269*	0.223	0.196	0.112	-0.495**	0.21	0.204	0.938**	1	
	Sig	0.334	0.328	0.201	0.608	0.263	0.377	0.716	0.048	0.031	0.076	0.121	0.376	0.000	0.096	0.106	0.000	.	
Text	r	0.078	0.029	0.067	-0.031	-0.093	-0.24	-0.225	-0.069	0.242	0.17	0.149	0.088	-0.025	0.194	0.189	0.302*	0.227	1
	Sig	0.542	0.82	0.597	0.811	0.471	0.056	0.074	0.593	0.054	0.18	0.24	0.491	0.842	0.124	0.135	0.015	0.072	.

**Correlation is significant at the 0.01 level (2-tailed).

*Correlation is significant at the 0.05 level (2-tailed).

Table 6.15. *Correlation matrix for studied variables (Comfort Meter measurements, MyLeaf measurements and Right-Now survey responses) (N=64). For a color version of this table, see www.iste.co.uk/kolokotsa/smartbuildings2.zip*

6.7.8. *Comparison between objective and subjective thermal sensation measurements*

Measurements with the Comfort Meter show increased operative temperatures and a slightly warm thermal sensation at all times in the RforI office space. Subjective measurements, also show a large number of slightly warm votes, however the tendency is for occupants to feel more neutral. In fact, even a slightly cool thermal sensation has been voted in some cases through the survey for the RforI office space.

In order to ensure that the concurrent Comfort Meter and survey responses are compared against each other, only the time-matched measurements were considered in this section. Even with comparison of concurrent Comfort Meter measurements and Right-Now survey responses, a discrepancy in thermal sensation is still observed (Figure 6.26). The plot reveals a trend for measured PMV votes to decrease as the day progresses, but they all remain very close or over +1. Such a trend is not observed for the subjective votes. Also, a large number of neutral votes is observed for the questionnaire measurements and even some slightly cool votes.

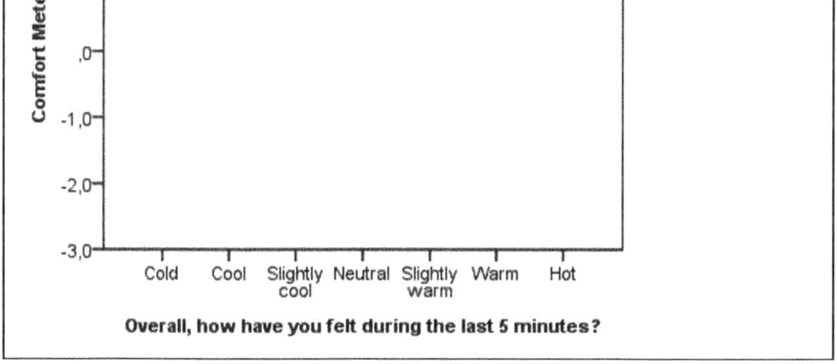

Figure 6.26. *Objective and subjective concurrent measurements of thermal sensation (PMV)*

It can be said that maybe the respondents feeling more cool or neutral are seated away from the Comfort Meter. Information about the exact location of the respondents was not collected for the purposes of ensuring the anonymity of the respondents and reducing the length of the survey.

Still, the reasons for the discrepancy between the Comfort Meter measurements and the MyLeaf and questionnaire survey findings should be investigated. The primary activity to be checked in this process is the calibration of the monitoring equipment (Comfort Meter and MyLeaf sensors).

6.7.9. *Determination of acceptable and unacceptable conditions*

In this study, dissatisfaction is considered to be the unacceptability with the thermal environment and the preference for change toward a cooler or warmer state. Average internal temperatures were grouped under these categories to help gain insight of the acceptable internal temperatures and the temperatures which the occupants of the Loccioni LeafLab feel uncomfortable in. Figure 6.27 illustrates the range of average internal temperatures for all studied zones, separating between satisfactory and unsatisfactory conditions.

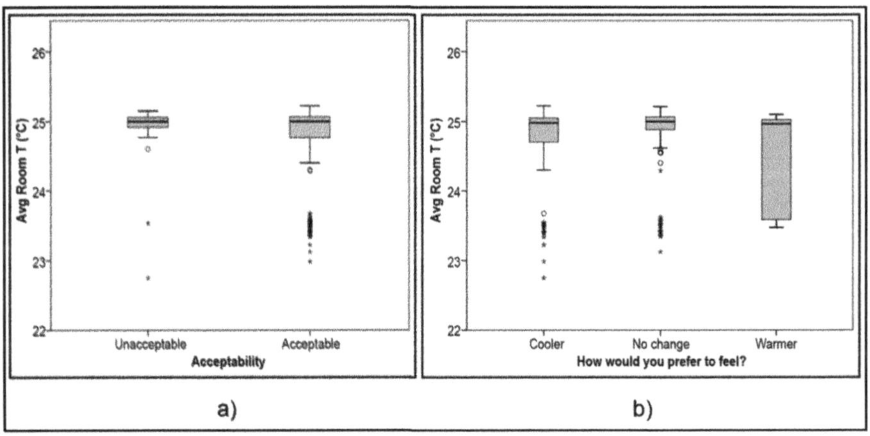

Figure 6.27. *Boxplots with average internal temperature vs. a) acceptability (left) and b) preference (right) (MyLeaf measurements)*

As observed through Figure 6.27 and Table 6.16, the mean acceptable temperature is only marginally lower than the unacceptable mean temperature. In both cases the temperature is between 24.5–25°C. However, the range of acceptable temperatures is broader than the range of unacceptable temperatures and spans towards lower values.

Figure 6.27 and Table 6.17 again suggest similar temperatures between those that would prefer for the thermal conditions to change and those that would prefer for them to remain the same. In fact, the mean temperature that respondents are satisfied in is marginally higher than the temperature under which they would prefer to be cooler. Also it is worth noting that the range of temperatures in which the respondents would prefer to feel warmer is much broader.

	Mean	Std. Dev.	N
Unacceptable	24.794	0.6102	19
Acceptable	24.728	0.5960	190

Table 6.16. *Acceptable and unacceptable temperatures*

	Mean	Std. Dev.	N
Cooler	24.693	0.6442	78
No change	24.790	0.5329	127
Warmer	24.385	0.7774	9

Table 6.17. *Preferred temperatures*

6.8. Conclusion

The thermal comfort in the Loccioni LeafLab office spaces was studied through subjective and objective measurements. The objective measurements involved the monitoring of environmental conditions through the building monitoring system (MyLeaf system) and through a portable Comfort Meter. Subjective measurements were carried out with an online questionnaire survey (Right-Now survey) that was answered three times per day (morning, midday and afternoon). The study of thermal comfort in the Leaf Lab office spaces had a duration of eight calendar days (July 22–29, 2016).

A total of 214 responses were collected with the Right-Now survey. About 51% of the respondents were from the RforI department to help support the more in depth study of the space with the Comfort Meter. From the other three departments the proportion of respondents out of the total was between 15% (Energy) and 18% (Industry). In Progettazione the proportion of respondents was 16%.

The objective and subjective measurements were matched based on the studied office space and the time of measurement. Descriptive and multivariate analysis was conducted on the collected data. The main findings of the analysis are as follows:

– Overall, no significant thermal comfort issue is noted in any of the LeafLab office spaces. In all individual spaces the mean thermal sensation was found to be in agreement with the prescribed ASHRAE range of $-0.5 < PMV < +0.5$ and the requirement for <10 PPD.

– The biggest proportion of respondents in all four spaces suggest neutral thermal sensation, however, a very large number of responses in RforI, Industry and Progettazione suggest a slightly warm thermal sensation as well. On the contrary, in the Energy department a large proportion of respondents were feeling slightly cool rather than slightly warm on many occasions. This is also reflected in the increased preference of the RforI, Industry, and Progettazione for a cooler environment.

– The most energy efficient way to satisfy the preference of respondents for a cooler environment would probably be through an increase in the air flow. Based on the Comfort Meter measurements, there is room to enhance air movement in the RforI department, since the mean air speed in the space was found to be 0.004 ± 0.0012 m/s.

– The mean acceptable temperature is only marginally lower than the unacceptable mean temperature. In both cases the temperature is between 24.5–25°C.

– The mean temperature that respondents are satisfied in is marginally higher than the temperature under which they would prefer to be cooler. Again they are both between 24.5–25°C.

7

Smart Meters and User Engagement in the Leaf House

7.1. Introduction

Smart metering systems add more information than conventional meters and offer many advantages such as cost effectiveness, accuracy and interactivity with users. Once in place, interactive smart meters can allow users to control and manage their individual preferences and consumption patterns, providing incentives for energy-efficient use through behavioral change.

User engagement and acceptance is a critical factor for the successful roll-out of smart meters. According to the Energy Efficiency Directive (EED) (Articles 9 and 12), appropriate advice and information should be given to users at the point of installation (Wilson *et al.* 2014). In particular, in line with Article 9, the Member States (MS) should ensure that the objectives of energy efficiency and benefits for final customers are fully taken into account when establishing the minimum functionalities of the meters. In addition, MS should ensure the security of the smart meters and data communication, and the privacy of the final users.

Smart meters can be a useful tool in giving customers up-to-date information on energy use and in enabling accurate billing. The advanced metering infrastructure (AMI) can reduce the environmental impact via lowering consumption and/or load (Valkering 2013). Nevertheless, to achieve a reduction in energy use the engagement of the user is crucial.

Chapter written by Niki GAITANI.

For a color version of all figures in this chapter, see www.iste.co.uk/kolokotsa/smartbuildings2.zip.

For the implementation of smart meters and the development of *efficient communication with the users*, it is necessary to carry out surveys to understand user needs and reveal possible problems or unclear issues regarding smart meters and their potential. The role of end-users in smart metering applications and their involvement in sustainable consumption is also acknowledged by Article 9 of the EED.

Current research has predominantly focused on the technical and system configurations involved with smart metering, with non-technological topics, such as consumer acceptance of smart metering systems, attracting considerably less attention from researchers (Torvinen and Ulkuniemi 2016).

Even though the importance of the end-user has been recognized by researchers and policy makers, there is lack of comprehensive knowledge on the engagement and confidence of the consumers (Giordano *et al.* 2013; Papantoniou *et al.* 2015).

This study aims to provide an insight into user perceptions on the smart metering services of the Leaf House, which is the first carbon neutral Italian house. In addition, the results will contribute to the design of consumer engagement with the smart grids of the future, bringing active demand technology to everyday life.

7.2. Methodology

The aim of this chapter is to provide an insight into user engagement in smart metering applications. Figure 7.1. illustrates the specific objectives.

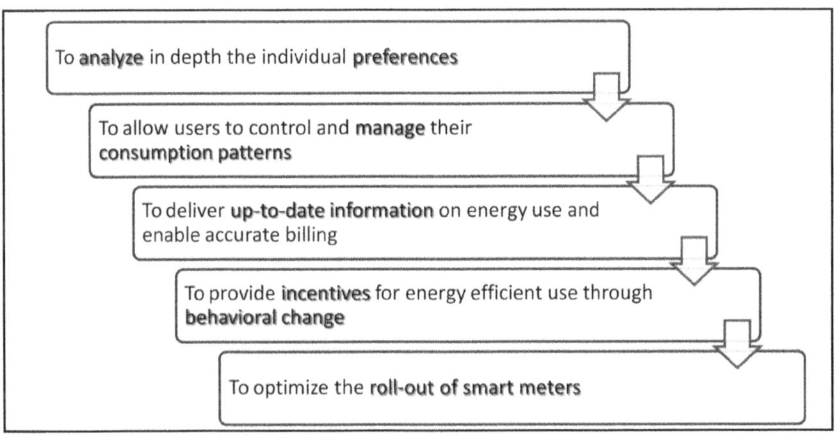

Figure 7.1. *Objectives of this study*

The *methodological procedure* consists of four main steps:

1) development of questionnaire;

2) users' survey and focus group;

3) data analysis;

4) results and discussion.

The data was obtained by a survey given to users of the *Leaf House*, a residential building which was designed and constructed with the view of minimizing energy consumption and emissions.

7.3. Analysis of user engagement

7.3.1. *Development of the questionnaire*

To fulfil quantitative data requirements, a questionnaire for the tenants/users is developed and applied to investigate the following topics:

– *Demographics, socioeconomics*: determines the basic socio-demographic and socio-economic characteristics of the sample such as: gender, average age, household size, income, and education.

– *Psychological, social and behavioral aspects*: deals with the consumption behaviors and attitudes of the tenants. It also identifies usual everyday energy and water behavior patterns and opportunities for promoting energy-efficient behavior change.

– *Information about environmental issues*: helps determine the existing (self-reported) knowledge on energy, water and environmental issues in general.

– *Health and comfort*: determines whether there are problems related to thermal comfort and indoor air quality, but also structural problems (condensation, dampness and mold) and health issues that should be considered in the design of the energy management plan.

– *Living situation*: gives an overview of the installed devices that consume electricity, and about the living situation of the participants.

7.3.1.1. *Set of questions*

1) Living situation and building:

– electricity;

– heating;

– domestic hot water.

2) Attitude toward energy saving:

– environmental domestic routine behavior;

– intention;

– personal norms and attitudes;

– habits.

3) Attitude toward technology.

4) Attitude toward nature.

5) Motives for participating.

6) Smart grid technologies:

– usage of variable tariffs;

– usage of devices;

– behavior;

– usage and evaluation of given digital information;

– comfort.

7) Standard demographic questions.

The questionnaire for tenants is designed to be answered by individuals. The goal is to recruit as many members as possible from each participating household who would independently complete their personal version of the survey. The proposed framework includes variables from psychological areas as well as general societal socio-demographics.

7.3.1.2. *Independent variables*

1) Socio-psychological variables: the following sets of behavior antecedents have been used:

– attitude (ATT);

– perceived behavioural control (PBC);

– personal norm (PN);

– awareness of consequences (AC);

– habits.

2) Knowledge

Responses are given on six-point scales, with scores ranging from 1 – "very well informed" to 5 – "very badly informed", and 6 – "don't know". Lower scores show higher levels of knowledge on environmental issues.

3) Socio-demographic variables: the following socio-demographic variables should be considered:

– income;

– education;

– occupation;

– age;

– gender;

– family status;

– household size.

Based on the feedback gathered, the template will be further fine-tuned to enhance its efficiency, usefulness and user-friendliness in the Smart GEMS and beyond the end of the project.

The questionnaire was based on the S3S 9 and ICE WISH 18 Projects. The questions were adapted to be in line with the Leaf House services and with the aim of user engagement analysis.

7.3.2. *Leaf House case study*

The Leaf House, shown in Figures 7.2 to 7.5, is a carbon neutral house in Italy, aimed at being as energy independent as possible and designed so as not to have any impact on the environment. It is a building made up of six apartments located in Angeli di Rosora and currently inhabited by Loccioni employees and visitors. Moreover, the building is located in Via Petrarca 4, 60030 Angeli di Rosora (AN) (43°28'23.4"N, 13°04'02.1"E).

Figure 7.2. *The Leaf House*

Figure 7.3. *Location of the Leaf House (Google maps)*

Figure 7.4. *The Leaf House*

Figure 7.5. *The control room in the Leaf House*

Building envelope and systems

Climate zone	Mediterranean
Year of construction	2008
Building utilization i.e. residential, office, commercial, industrial, etc.	Residential apartment and garages for employees and for guests
Building occupancy schedule	It comprises 6 apartments: – 8 people living regularly in 4 of these apartments, – 2 guest apartments. Apartments occupied by employees: 18:30–08:00 Apartments occupied by guests occasionally for a few days from 18:30–08:00
Energy Utility	Enel
Energy tariffs (€/kWh)	0.06750 €/kWh – 08:00–19:00 working days during the year 0.05250€/kWh – 19:00–08:00 working days, weekends and official holidays[1]
Gross floor area (m^2)	Approx. 220 m^2 + 40 m^2 garage
Number of storeys	4
Architectural plans or 3D images of the building	The building has a north-west/south-east orientation.
Building envelope description (i.e. wall – roof – floor structure, type of façade, percentage of window to wall area, roof type, etc.)	The Leaf House is a residential building designed and constructed with the view of minimizing energy consumption and emissions. The structure has a rectangular plan and is spread over 4 floors, of which 3 are above ground on the north-side, because the house is close to a hill and the basement. The building has a reinforced concrete structure with brick-concrete secondary structure for walls. The building is south oriented and the ratio between the lengths of the south and east façades was set to 1.34 to maximize solar gains during the colder seasons. The building is located on a hill and so only three façades are completely exposed. The principal south-east façade is characterized by door-windows for the day zones and common windows for night zones. The lateral façades have a minor number of windows with regular sizes for the secondary bedrooms. In the rear part of the house facing north, the sunlight arrives through solar tubes. The total percentage of window is equal to 4.25%.

1 https://www.enelservizioelettrico.it/it-IT/tariffe/uso-domestico/biorarie/d3.

	The wooden roof has a rounded free form that recalls the shape of a leaf and makes up the loft of the apartments of the top floor. The roof, the solar thermal panels and the balcony have been designed to behave as solar shadings during the hottest months. The floors are brick–concrete (20 + 4 cm thickness) and made of both wood and terracotta, natural and comfortable materials similar to what would traditionally be used and to the natural dimension of living. The walls structure is composed of an external layer of EPS with a thermal conductivity factor of 0.036 W/mK and a thickness of 18 cm; it insulates more or less just like a thickness of about 1.5 m of bricks. 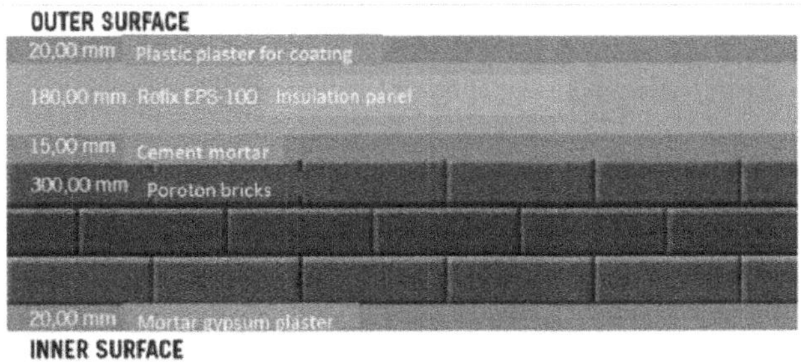 The attention paid to heat bridges is also remarkable: fixtures, rolling shutters, windows, doors and the relative structures must guarantee maximum insulation. This is not only a benefit for extraordinary thermal insulation but also acoustic insulation up to 43 dB. The Leaf House flats are absolutely noisefree. Conceived as a prototype house, to meet the liveability needs in a domestic environment, compatible with energy saving, inside there are networks of systems and sensors managed by a home automation system that allows control and integrated management. Through a display, the home automation system enables tenants to control and switch on/off the lights in the apartments.
HVAC System (kW)	– Geothermal heat pump – Hot water boiler – Air handling unit Air ventilation: outer air is heated in winter and cooled in summer. In addition, it is naturally pre-conditioned through an underground path of about 10 m before reaching the AHU. Radiant floors in each apartment for heating and cooling
Lighting System (kW)	Fluorescent lamps and LEDs
DHW System (kW)	-

Energy Management and Control System	About 1,000 sensors installed: – indoor comfort (temperature, humidity, CO_2), – thermal and electrical measures, – energy consumption, – rain water level. Building automation: each apartment has a display that enables tenants to: – set the desired temperature in the apartment, – monitor lights, – check if windows/doors are open, – see energy consumptions.
Current state of repair and year of last major renovation of building envelope and main technical installations respectively	Many changes and upgrades have been made over the years and this affects the annual energy consumption. In 2012 a 5.8 kWh electric storage was introduced in the Leaf House and it operated until late 2014: this caused an increase of the energy self-consumption due to the storage of PV production.

Energy data

Energy consumption for 2015	61,597 kWh
Energy produced in 2015	23,076 kWh
Energy from the Leaf House to the grid in 2015	7,057 kWh
Measured average annual consumption of electrical energy and other type of fuels over the last 1 year (kWh)	5,133 kWh
Peak Load (kW)	860 kW (on April, 2015)

Energy generation and storage data

Energy generation by RES	PVs 20 kWp / 150m^2

7.3.2.1. *Leaf House tenants' focus group*

The majority of the tenants of the Leaf House, current and previous, have joined the focus group: through emails, discussions and meetings the remarks obtained could be summed up as:

– there is a need for a more *interactive monitoring system*, so the inhabitant may more easily control parameters related to comfort and energy performance;

– new tools are needed with *indications* for the users when they consume more;

– more *customized options* for the everyday user should be provided;

– the *operative temperature* should be used as the optimum set point;

– the *thermal inertia* of the Leaf House building results in a slow response to a change of indoor temperature;

– the behavior of the users is linked to the general profiles of the inhabitants, cultural aspects, and cost and environmental benefits;

– the motivation should be connected to the values of sustainability;

– the Leaf House may provide an *educational environment* as a Living Lab.

The key points that have been addressed by the focus group are illustrated in Figure 7.6.

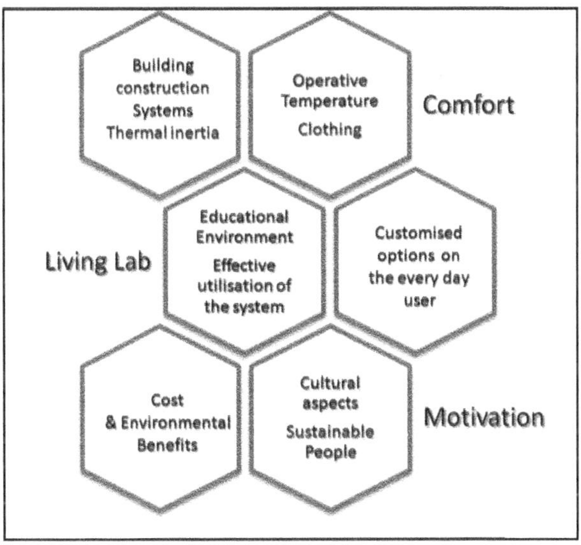

Figure 7.6. *Summary of the main topics discussed at the focus group*

7.4. Results

All tenants of the Leaf House were invited and participated in the survey by filling in the questionnaire developed in this chapter and that covers the areas described in Figure 7.7.

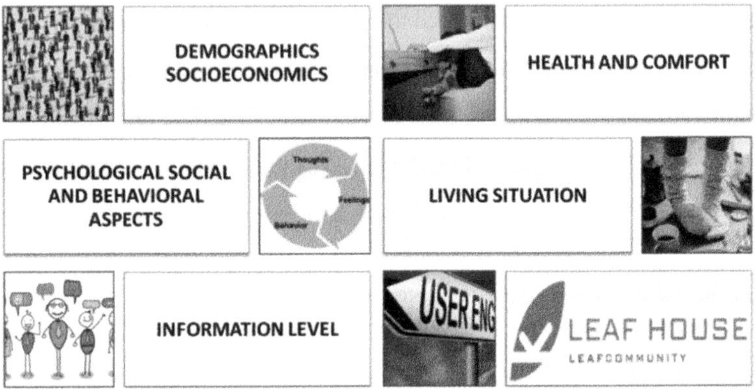

Figure 7.7. *The topics covered in the questionnaire*

The results obtained from the analysis of the questionnaires are presented grouped into these five topics.

7.4.1. *Demographics, socioeconomics*

Within every questionnaire the so called standard demographic questions are crucial. The information about age, gender, income, education and occupation are the most basic information about the end user (Figure 7.8).

Finally, in this part of the questionnaire it is underlined that:

1) 95% of the tenants are 25–49 and 5% 50–64 years old; it should be noticed that guests have been included in the analysis too;

2) 70% of the tenants spend less than half of the day at home and are mostly out of the home, while 15% stay at home only for sleeping at night, 10% stay at home most of the day and are only out of the home for short periods, and 5% stay at home longer than half of the day and are out for a few hours;

3) 63% of the tenants have an education level at the first tertiary stage (University degree, MSc) and 37% hold a PhD;

4) 34% of the tenants have a monthly income between 1,001 and 2,000 Euro, 33% between 2,001 and 3,000 Euro, 11% between 3,001 and 4,000 Euro and 22% n/a;

5) the great majority of tenants are employed full time.

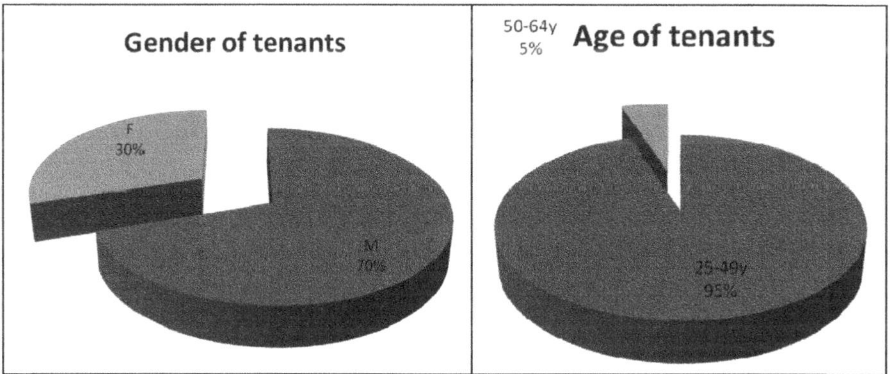

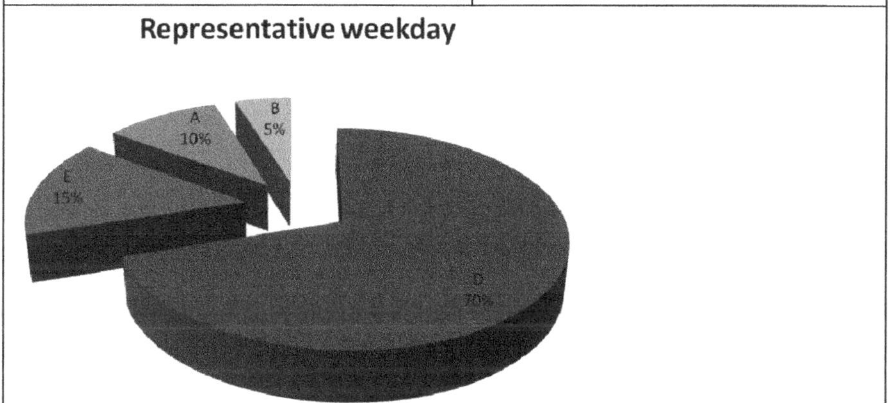

A. at home most of the day and only out of home for short periods
B. at home longer than half a day and out for a few hours
C. at home half a day and out of home half a day
D. less than half a day at home and mostly out of home
E. at home only for sleeping at night

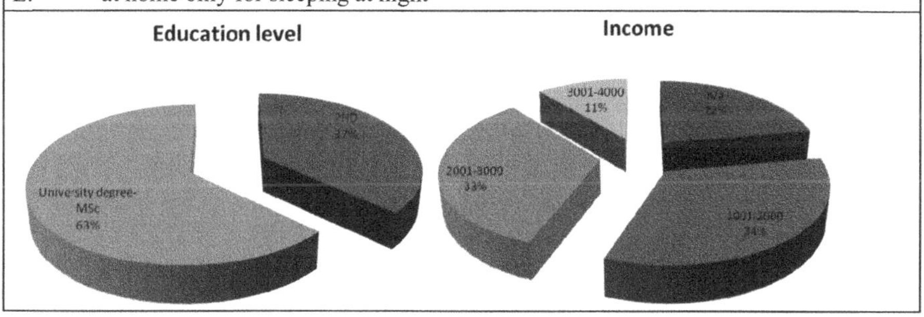

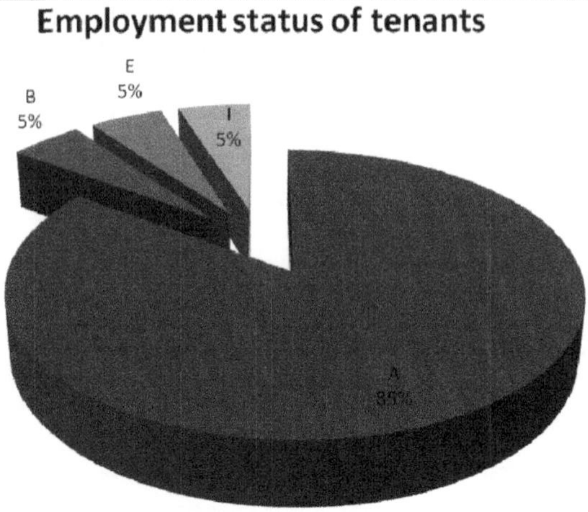

A. Employed full-time
B. Employed part-time
C. Self-employed/working from home
D. Retired
E. Full-time student
F. Unemployed
G. Temporarily sick or injured
H. Long-term sick, injured or disabled
I. Other

Figure 7.8. *Demographics, socioeconomic background*

7.4.2. *Physiological, social and behavioral aspects*

The questions related to psychological and behavioral aspects have revealed results as shown in Figure 7.9:

– tenants know that the heating is on during winter but their answers differ as to which months this covers;

– 78% of the tenants replied that they use central heating, 11% separate heating devices and 11% that they don’t know;

– there is a variation regarding the hours of heating, 63% of the tenants replied that they don’t know;

– 55% of the tenants replied that they put on more clothes when they feel cold and 27% select to increase the temperature;

– the majority replied that they have a cooling device;

– according to the feedback, 50% of the tenants know that the cooling is on during the summer but their answers differ as to which months this covers;

– 56% of the tenants replied that they don't know the hours of cooling per day;

– 46% of the tenants replied that in order to keep cool they increase the temperature, 23% tend to wear fewer clothes, 15% open the window, and the remainder call a service or use another cooling device;

– 89% of the tenants replied that the climate in their home is being monitored and 11% that they don't know if it is;

– the majority of tenants do not know the tariffs.

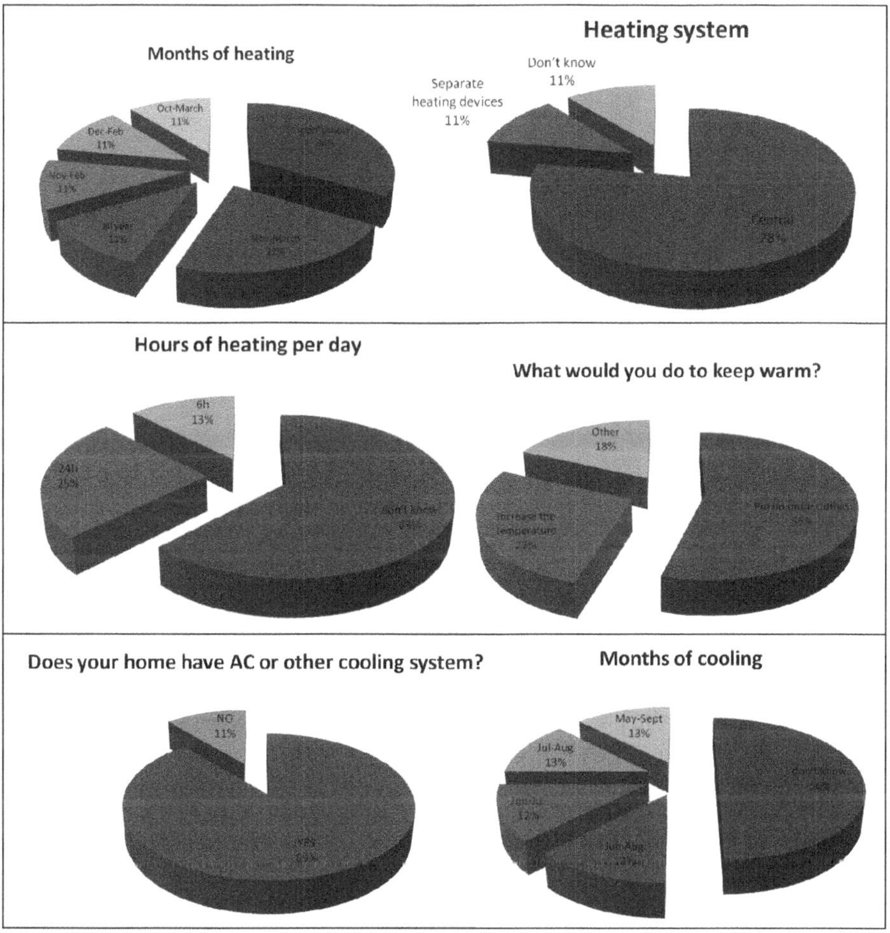

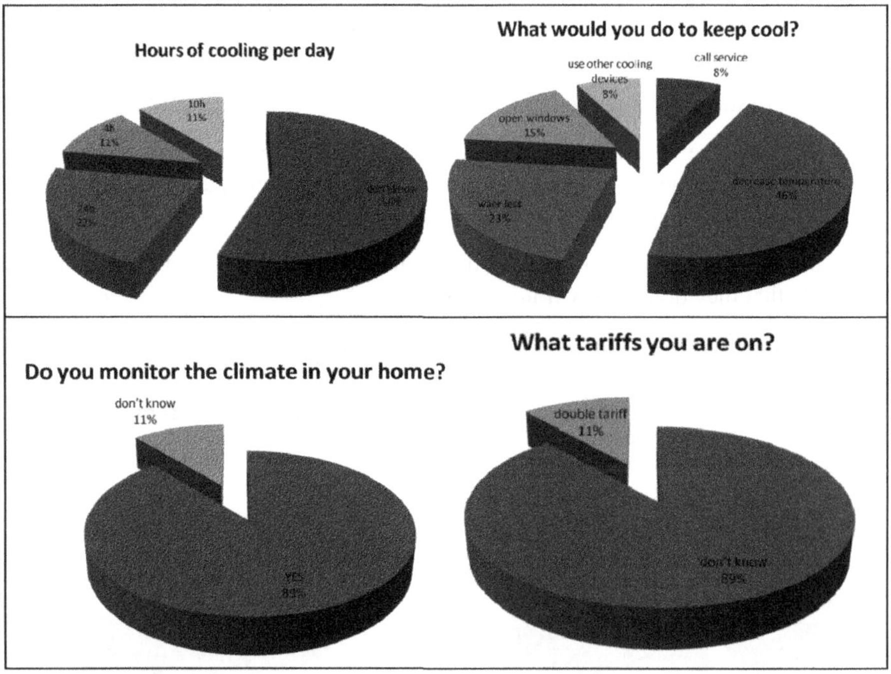

Figure 7.9. *Physiological, social and behavioral aspects*

7.4.3. *Information level*

Overall, according to the tenants' responses regarding the information level (Figure 7.10):

– 78% of the tenants feel fairly well informed about the possibilities for protecting the environment;

– 56% of the tenants feel fairly well informed about the consumption of energy and water in their home;

– with regard to the possibilities of saving energy and water in their home, 56% of participants feel fairly well informed, 33% feel neither well nor badly informed, and 11% do not know.

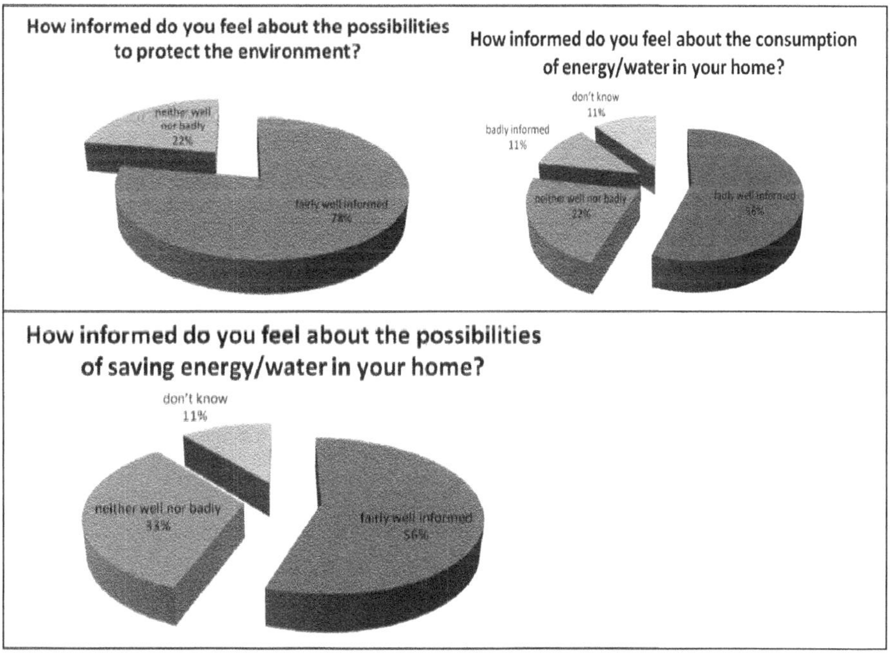

Figure 7.10. *Information level*

7.4.4. *Health and comfort*

With respect to health and comfort (Figure 7.11):

– condensation on windows/walls/ceilings has been indicated as a problem in bathroom (22%), kitchen (11%) and living room (11%);

– 78% of the tenants feel thermal comfort indoors during the heating period;

– the majority feel thermal comfort indoors during the cooling period;

– 56% of the participants do not have any health issues, 33% suffer from allergies, and 11% note circulatory problems;

– within the scale 1 (satisfactory) to 5 (unsatisfactory), the majority of the tenants is satisfied (1 and 2) with the overall comfort during winter;

– within the scale 1 (satisfactory) to 5 (unsatisfactory), the majority of the tenants is satisfied (1 and 2) with the overall comfort during summer.

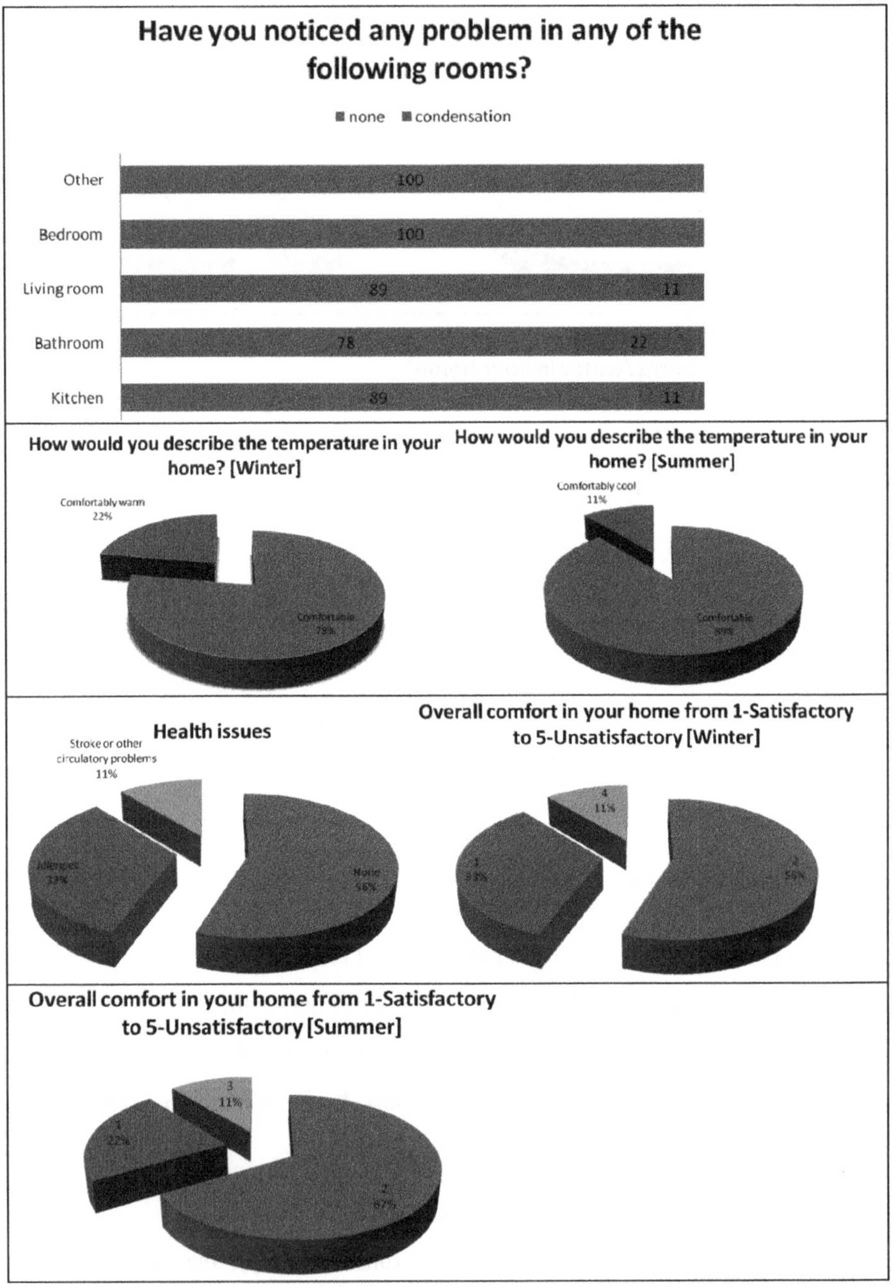

Figure 7.11. *Health and comfort*

7.4.5. *Living situation*

Regarding the living situation (Figure 7.12):

– all the tenants use the fridge, the computer and the TV and usually the oven, the washing machine and the freezer;

– 80% of the tenants keep the same temperature in the bedrooms;

– 45% of the users never reduce the temperature manually, 33% of the tenants change the temperature manually sometimes, 11% rarely and 11% always;

– to freshen up the air, the tenants usually select natural ventilation by opening the windows;

– 2/3 of the tenants take a shower five times per week for about 5–10 min.

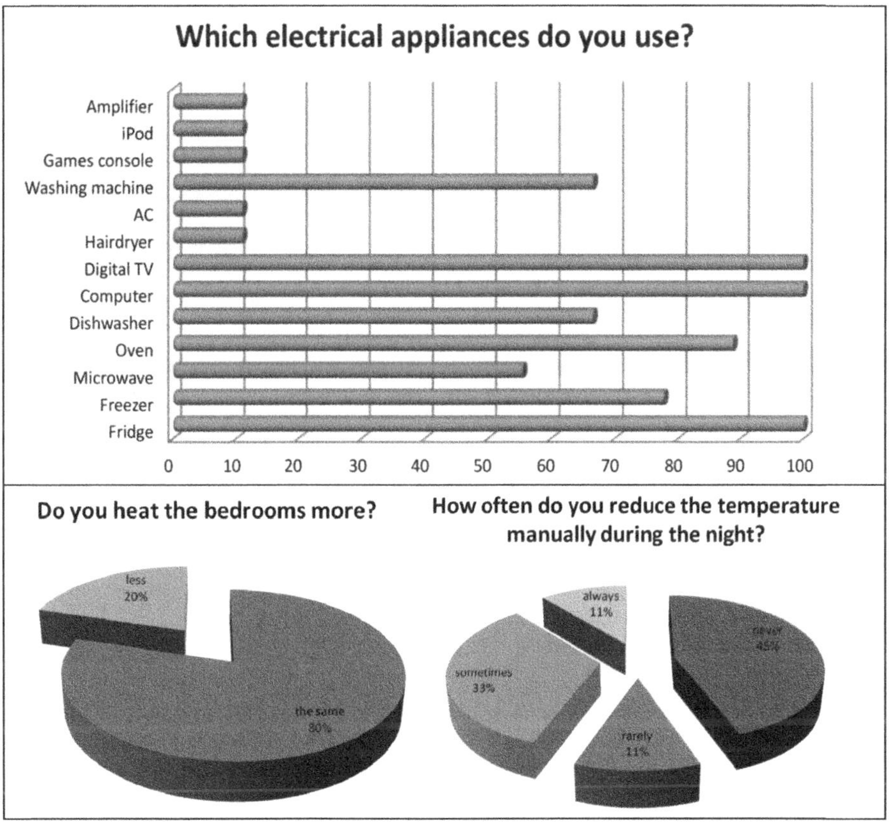

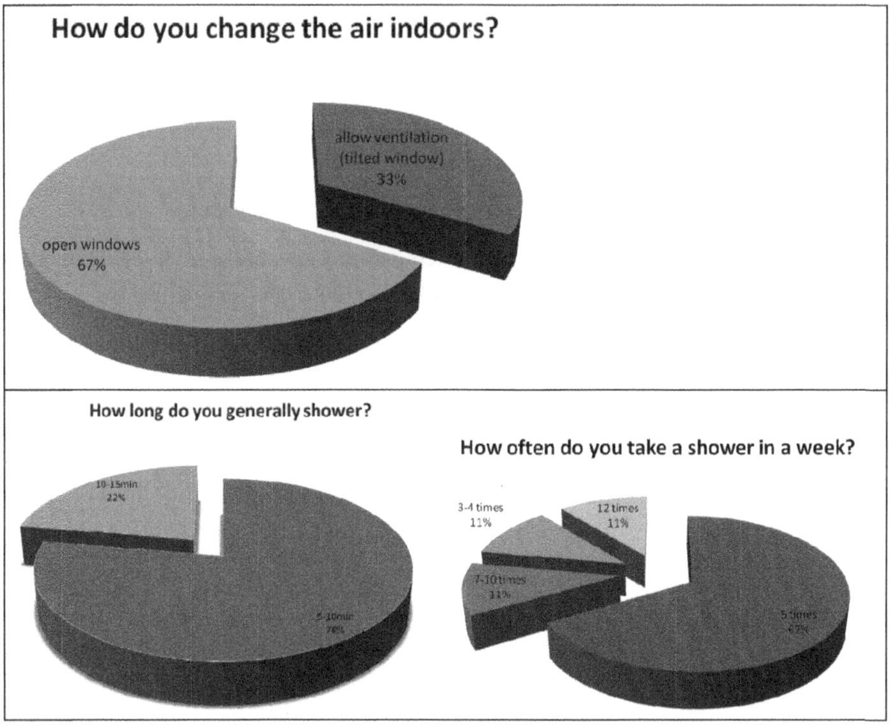

Figure 7.12. *Living situation*

7.5. Conclusion

The terms user engagement and user confidence emerge as key factors for smart grid success. The results highlight activities that can be summarized into two main points.

1) Understanding the user:

– the users of the house are overall satisfied with the living conditions in the building;

– however, the majority of the tenants do not feel adequately informed about their consumption and the possibilities to save energy and water.

2) Engagement of the user:

– the user interface of the monitoring system could be further improved to provide information about energy consumption and customized options for the everyday user;

– it is important to inform and advise the end-users about the effective utilization of the metering system and the energy saving potential;

– utilities should inform the end-users about the benefits of smart meters to motivate news patterns of energy use.

To achieve economically rational consumers, smart grid designs must look beyond strict technological aspects and recognize that a smart user who is actively engaged with energy is decisive in what is proposed by demand side management. The systems need to be designed for real, as opposed to ideal, users. Smart grids are a key feature of energy performance, with the goal of better aligning energy generation and demand.

8

Integration of Energy Storage in Smart Communities and Smart Grids

Smart grids are one of the major challenges of the energy sector for both the energy demand and energy supply of cities and communities (Kolokotsa *et al.* 2016b; Colak *et al.* 2016). A transition from traditional electricity grids where energy is produced centrally and distributed to the various energy consumers is necessary to exploit distributed energy sources, power loads, and storage components. Traditional grids lack flexibility in power generation and load operation, thus forming a semi-autonomous entity with limited energy management capabilities (Erlinghagen and Markard 2012). Moreover, a smart micro-grid can operate connected to the main grid or in island mode.

The deployment of smart grids asks for innovative techniques to predict and control electricity requests to the grid, in order to reduce peaks, improve grid stability, and optimize electricity costs (Lund *et al.* 2012). Smart grids can even contain millions of smart meters, which produce a huge amount of data concerning energy consumption. In addition to these time series data, additional correlated information about energy sources, weather, occupancy, building performance, and so on can be provided (Javed *et al.* 2012).

Chapter written by Denia KOLOKOTSA, Nikos KAMPELIS, Angeliki MAVRIGIANNAKI, Marco GENTILOZZI, Filippo PAREDES, Fabio Maria MONTAGNINO and Luca VENEZIA. This chapter is based on a previous work by the authors, "On the integration of the energy storage in smart grids: Technologies and applications", *Energy Storage*, 1, 2019, reproduced with permission from Wiley.

For a color version of all figures in this chapter, see www.iste.co.uk/kolokotsa/smartbuildings2.zip.

The actions toward a fast implementation of the smart grid has increased recently due to policy and regulatory initiatives that promote smart grid reliability, integration of renewables, demand response, and energy storage (Moslehi and Kumar 2010; Siano 2014).

Energy storage is a critical component for smart grids (Carpinelli *et al.* 2013; Saffari *et al.* 2018; Duffy *et al.* 2019). The significance of EES in energy networks is analyzed in Das *et al.* (2018). Optimal sizing and operation of storage systems provides peak power demand effective management, allows the maximization of renewables' penetration, supports power quality, and reduces energy networks' expansion needs (Psola *et al.* 2014; Yang and Hsiung 2018). Moreover, the increased capital and operational costs to manage peak demands, the requirements for grid stability, resilience, and reliability have increased the interest of storage for grids (Dunn *et al.* 2011).

Typical energy storage systems for smart grids include (Chen *et al.* 2009; I.E.C. 2011):

– electrical energy storage, that is, electrostatic energy storage including capacitors, super capacitors, and magnetic/current energy storage;

– mechanical energy storage, such as flywheels, compressed air, and hydro-pump energy storage;

– chemical energy storage, such as electrochemical conventional batteries, fuel cells, solar fuels, metal-air batteries and thermochemical reversible processes;

– thermal (sensible, latent and thermochemical) energy storage.

Energy storage may also be provided through the control of building thermal storage (Thatte and Xie 2012; Kohlhepp and Hagenmeyer 2017) and frequency control on loads.

The various categories of energy storage systems are depicted in Figure 8.1.

Much effort has been made to analyze the role of storage systems in the future development of smart grids and their interconnection in building, community, and city scales (Koohi-Kamali *et al.* 2013; Roberts and Sandberg 2011; Pérez-Romero *et al.* 2013; Espe *et al.* 2018). Smart grid implementation, zero energy and carbon community targets, distributed energy generation, and increased penetration of renewables are mainly influenced by the successful integration of energy storage systems (Karlessi *et al.* 2017). The EES in power systems and utilities infrastructure stability and quality is also an important aspect and focus of research for various researchers (Koohi-Kamali *et al.* 2013).

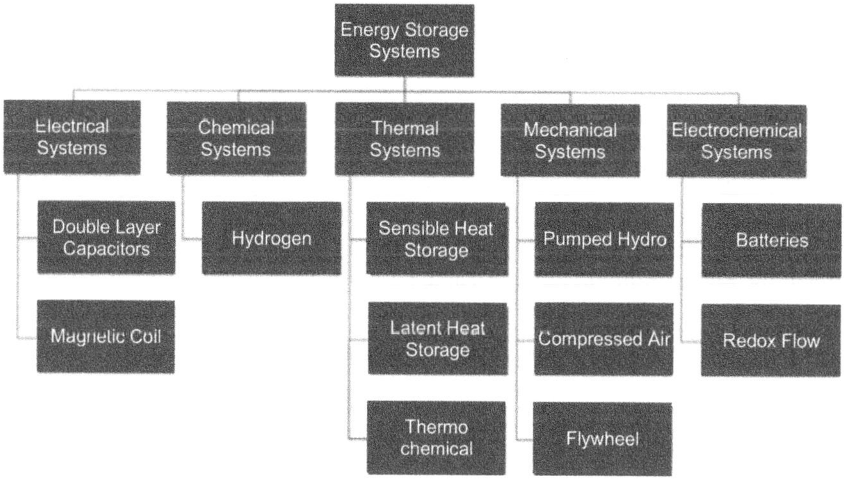

Figure 8.1. *The energy storage systems categories*

Another aspect of research includes the optimized operation of storage systems in smart grids. Some examples of optimization methodologies for energy storage in smart grids include the Particle Swarm Optimization for scheduling the operation of various energy resources and electric vehicles (EVs) (Hutson *et al.* 2008; Soares *et al.* 2012), as well as the use of genetic algorithms and multi-objective optimization (Siano 2014; Bahmani-Firouzi and Azizipanah-Abarghooee 2014).

The integration of storage in smart grids requires extensive research in analyzing performances for real case studies, setting key performance indicators, and benchmarking. To this end, the aim of the present work is to review the different research approaches on the actual implementation of storage systems when coupled with smart grids. The chapter starts with an analysis of the various energy storage systems and their performance when coupled with smart grids. Two case studies covering the actual implementation and results of electric and thermal energy storage are examined in section 8.2, while section 8.3 outlines some conclusions and future prospects.

8.1. Energy storage systems in smart grids

8.1.1. *Electrical and electrochemical energy storage in smart grids*

Electrical energy storage systems (EES) are very important for smart grids. Their role generally depends on the number of charging-discharging cycles and the duration of their operation (IEC 2011).

The benefits of electrical energy storage in grids are:

– Time and peak shifting: for power utilities generation cost can be reduced if energy is stored during off-peak hours and released during peak hours.

– Power quality: power utilities should supply power voltage and frequency within specific limits. EES can perform frequency regulation and this is done by adjusting the supply to the changing demand (Lucas and Chondrogiannis 2016; Mercier *et al.* 2009). In isolated power grids, electrical storage systems support the power utility to provide stable power to the customers.

– End-user peak shaving, load shifting, and dynamic pricing: energy storage can be used by customers, such as industrial users for peak shaving, in order to minimize the part of their energy bill that varies according to their highest power demand. Such a service is very profitable if the peaks are sufficiently predictable and of a relatively short duration (Graditi *et al.* 2016; Parvania *et al.* 2014; Khan *et al.* 2016).

– Continuity of energy supply: a storage device is able to substitute the network in case of interruption. EES are often used as an emergency power supply in case of outage (Fedak *et al.* 2014).

– Limitation of disturbances: the storage can help customers to comply with commitments related to limited disturbances thus improving grids' resilience (Moslehi and Kumar 2010; Farraj *et al.* 2018; Ayar *et al.* 2017).

– Compensation of the reactive power: an electrical storage system is able to locally compensate for the reactive power via the power electronics converter (Maleki *et al.* 2017; Utkarsh *et al.* 2019).

Batteries are the first EES to be considered in smart grids. Battery systems can offer a number of high-value opportunities, provided that lower costs can be obtained. The most widely used batteries are lead acid nickel-cadmium, lithium, and sodium sulfur (Shukla *et al.* 2001; May *et al.* 2018; Roberts 2008).

The main concerns for the wide adoption of these batteries are the overall cost, limited number of charging cycles (or lifetime), depth of discharge, low energy density, and sustainability of the materials used.

On the other hand, flow batteries can be an alternative. For example, Vanadium redox flow batteries (VRFB) are a promising option to mitigate many of these shortcomings (Alotto 2014). Lucas and Chondrogiannis (2016) demonstrated how a VRFB based storage device can provide frequency regulation and peak-shaving functions.

A review of sodium-sulfur batteries, redox-flow batteries, and lithium-ion batteries can be found in Dunn *et al.* (2011).

Moving to another subcategory of electrical energy storage, ultra-capacitors are another emerging solution. Ultra-capacitors (Atmaja and Amin 2015):

– have an almost 20 times higher capacitance than normal capacitors;

– have a longer lifecycle;

– are insensitive to environmental variations in temperature.

The ultra-capacitors can be applied for:

– optimized operation of renewable energy sources (Singo *et al.* 2008);

– EVs power management (Atmaja and Amin 2015);

– hybrid energy storage systems (HESS) (Smith and Sen 2008).

Superconducting magnetic energy storage (SMES) is an alternative for high density storage systems' requirements (Colmenar-Santos *et al.* 2018). In a case study for Malaga, Spain, SMES are used to eliminate grid signal fluctuations created by industrial clients. The specific industrial clients incorporate three-phase induction motors that deteriorate the quality of the grid signal. Also, hybrid configurations of SMES, for example, SMES with batteries, with fuel cells, and so on are very encouraging solutions.

Furthermore, EVs are increasingly being considered as necessary technologies for energy storage in smart grids. EVs can operate in two different modes that can serve as energy storage in smart grids (Pang *et al.* 2014):

– "grid-to-vehicle" (G2V) operation, where the vehicles are charged using electricity from an electric power grid.

– "vehicle-to-grid" (V2G) operation, where vehicles are discharged to an electric power grid during parking hours. Another approach is the integration and interconnection of EVs to specific buildings, formulating the "vehicle-to-building" (V2B) alternative (Nefedov 2018), which allows buildings to become energy prosumers and increase the share of renewable resources.

Through these operational modes, EVs can act either as a load or as decentralized energy storage in a smart grid. The expansion of EVs includes significant challenges concerning grid stability. Smart bidirectional charging strategies, as well as management strategies of EVs with RES have been proposed by various researchers (Rahbari *et al.* 2017; Mozafar *et al.* 2017; Mukherjee and Gupta 2015; Luo *et al.* 2016). Accurate information regarding the state of the EVs connected to the electrical grid is required. Examples of EVs case studies are tabulated in Table 5.1. A more detailed implementation of EVs in various countries can be found in E. A. International (2012).

Case study	Characteristics	Main results	Ref
Electric buses, Guwahati city, Assam, India	A high capacity energy storage device is used between the grid and transportation system. A solar plant is introduced to curtail the system dependence on the grid.	Reduction of internal combustion engine vehicles and improved air quality. Electrification of transportation with support from renewable energy. Transformation of transportation.	(Bhaskar Naik *et al.* 2019)
Application of electric vehicles (EVs) on the Caribbean island of Barbados	Peak electricity demand 167.5 MW and 240 MW installed generation. 10 MW utility-scale solar Photovoltaics (PV) and 20 MW of distributed solar PV on residential and commercial rooftops. The EVs are either considered as static demand or smart charging with V2G capability.	Decrease of grid-connected storage up to a 13% if EVs are charging during the day. V2G allows increased penetration of renewables and reduces the investment needs in grid-connected storage 12% in case of EV night charging and to 20% for day charging.	(Taibi *et al.* 2018)
São Miguel Island, Azores, Portugal	Medium voltage distribution network is analyzed. A transformer is utilized, which supplies one private industry client via a 250 kVA, 10 kV/0.4 kV oil-immersed transformer. An EV charging scheduler is analyzed	With 60% EV penetration, the loss of life of the transformer is reduced by a range of 7.54% at 60% EV penetration to 76% at 100% EV penetration. With the EV charging scheduler, a 98% of loss of life reduction is achieved.	(Godina *et al.* 2016)
Tenerife, Spain	Data from the cars, mobility and power demand have been taken into account to analyze the EVs penetration.	Considering various levels of penetration to the grid, EVs are utilized as storage systems that allow the significant reduction of the amplitude difference between valleys and peaks of the electric energy demands curve, and thereby contribute to the efficient management of smart grids.	(Colmenar-Santos *et al.* 2017)
Korcula, Croatia	Analysis of 100% renewable energy sources (solar and wind) in combination with 100% share of vehicles is performed.	The import and export peak loads have not been affected by the smart charge share reduction of EVs.	(Dorotić *et al.* 2019)

Case study	Characteristics	Main results	Ref
UCLA campus, city of Santa Monica	UCLA Smart Grid Energy Research Center (SMERC) is used as a testbed. Two stages of optimization are performed: (1) to reveal the optimal EV load profile for the day-ahead (DA) energy market; (2) to control all online EVs in the system to follow the optimal load profile and charging scheme generated by stage 1. EV user patterns are analyzed to optimize the EVs operation and couple it with a demand response program.	An 18% energy cost reduction is achieved while lowering the computational cost of the control system.	(Xiong *et al.* 2018)

Table 8.1. *Electric vehicles case studies*

Combinations of the aforementioned technologies provide significant advantages. For example:

– The use of battery packs that are incorporated in EVs equipped with bidirectional charging systems was proposed by Uddin *et al.* (2017). Commercially available lithium-ion batteries (C6/LiNiCoAlO2 type) are used and a degradation model is developed and validated. This model is integrated into a smart grid management algorithm. It is shown that the smart grid is able to extend the life of the EV battery. The EVs' battery pack capacity fade is reduced by up to 9.1% and its power fade is reduced by up to 12.1%.

– HESS with a high energy storage battery bank and ultra-capacitor unit was proposed by Aktas *et al.* (2018) to support the integration of PV in electricity grid. During load changing, the ultra-capacitor group is discharged and then the battery group supplies the load. Thus, battery life is preserved by reducing the deep-discharge rate.

– The integration of (1) flywheels; (2) plug-in electric vehicles (PEVs); and (3) price responsive thermal loads is presented in Thatte and Xie (2012). The storage is used for the day-ahead market using dynamic management systems and revenue obtained by the energy storage over a duration of one day, in $/MW, is used as a basis of comparison between different storage technologies (Kolokotsa *et al.* 2009).

8.1.2. *Mechanical energy storage in smart grids*

In mechanical systems, energy storage is well established in the form of large-scale hydro-pumped systems (HPS), but also medium- and smaller-scale systems (Carpinelli *et al.* 2019). Various examples show the performance of HPS in smart grids. For example:

– Chaudhary and Rizwan (2018) provide a typical example of using HPS in smart grids. During off-peak hours or surplus electricity generation by the PV, the water is pumped to the upper reservoir. During peak hours or lower PV generation, the water flows from the upper reservoir to the lower through the turbine/generator to produce electricity and supply power to loads. The HPS includes a turbine unit and motor. A variable speed doubly fed induction machine is considered.

– The incorporation of HPS in Sardinia is discussed by Ghiani *et al.* (2014). Sardinia has limited interconnections with the mainland and high solar and wind energy resource potential. On the island, the power system can be seen as a microgrid. HPS is used for load balancing. During off-peak hours, power is utilized to run the pumps, whereas during on-peak hours the stored water is released through turbines to produce electric power. This increases the renewable energy production penetration.

– The electrification of Southern Cameroon is analyzed by Muh and Tabet (2019). Nine alternative scenarios are investigated including hybrid renewable energy sources and energy storage. The scenario that combines PV, diesel, small hydro, and battery system was found to be the one with an optimum cost, with a cost of energy equal to 0.443 $/kWh, which is much lower than the renewable only system or the PV, LPG, and battery system.

– A residential settlement is exploiting PV with pico-hydro turbines and HPS to minimize the energy cost in a dynamic pricing electricity system (Anilkumar *et al.* 2017). During periods when the energy price is low, the energy produced by the PV is stored as gravitational energy to the pump storage and is used by the pico-hydro turbines when the energy price is high. A case study in Tiruchirappalli, India is analyzed.

In addition, compressed air energy storage (CAES) can be considered as one storage technology that has a very high ESOI index. An ESOI index is the total amount of energy stored over the lifetime of a storage technology unit, divided by the amount of energy used in producing that unit (Cazzaniga *et al.* 2017). An example using CAES is described in Cheng and Choobineh (2016). In this research CAES is combined with a wind energy conversion system to provide distributed energy resources and storage to a microgrid, in order to minimize its dependence on the transmission line in case of the microgrid's expansion. This integration can significantly decrease the demand on the transmission line, while simultaneously reducing the community's energy cost by 18%. Moreover, CAES can moderate fluctuations of renewable energy production due to the large-scale storage capacities, high ramp rate, and quick start-up time, and are therefore an attractive technology for smart grids (Zhang *et al.* 2016).

Although CAES is a mature and attractive technology for grid energy storage systems there are shortcomings that need to be considered. The CAES efficiency is much lower than similar technologies such as HPS, attributed to the fact that during compression, thermal energy is wasted. Thermal energy storage coupled with CAES is proposed to overcome the aforementioned issue (Ortega-Fernández *et al.* 2017). In this framework, the thermal energy produced in the compression stage is stored in a TES unit for its subsequent deployment during the expansion stage, realizing an Adiabatic-CAES plant with an efficiency that reaches almost 70–90% depending on the configuration (Hartmann *et al.* 2012). Another CAES application shows that the specific energy storage system can balance fluctuations in power generation and frequencies instabilities. An air flow controller is utilized and designed to control the air flow from the CAES system, so that the microgrid follows the various load demands to maintain a stable frequency (Latha *et al.* 2012).

Some examples of mechanical storage systems utilized in grid applications are tabulated in Table 8.2.

Case study	Characteristics	Main results	Ref
Flywheel sizing study for homes, Austin, Texas	A flywheel energy storage sizing study is performed using home energy data for 741 homes at the Mueller Development Community in Austin. The energy data include home energy consumption and solar photovoltaics (PV) generation. 25% of homes have a 6 kW PV. The trade-offs between energy storage capacity and power requirements are assessed at the home, transformer, and community levels.	At the community level diurnal energy storage requires 1,200 kW flywheels, while for power smoothing 300 kW are sufficient. The peak power in the power smoothing configuration is reduced by 8% for the whole community.	(Hearn *et al.* 2013)
CAES application in Clemson University, South Carolina	A solar assisted microgrid is developed using PVs 15 kW. Compressed air energy storage is utilized to supply a MicroTurbine and facilitate its operation.	The average power generation on a daily basis by the PV panels runs a 5.2 kW CAES for 5 hours, storing 1108 kg of air at a 1.2 MPa pressure. This corresponds to 8.1% efficiency improvement in energy generated by the system with CAES, compared to conventional production.	(Patel *et al.* 2014)

Integration of hydropumped systems (HPS) in Sardinia	The Sardinian power system is simulated under the hypothesis of considering a base-load of 1200 MW, and taking into account the variability of Wind and PV energy production. Sardinia is connected to the Italian mainland with a high voltage direct current (HVDC) connection called SAPEI, which allows 1000 MW import/export with the mainland. HPS is utilized to provide additional load at times of high RES generation and low electricity demand, enabling further RES peak capacity and/or regulating the export through HVDC connections.	HPS is used when the production overcomes the SAPEI limit of 1000 MW, and this is done for 395 hours per year. Also, the released energy stored from the HPS helps to reach the 1000 MW.	(Ghiani *et al.* 2014)
HPS of Alps as the battery of Europe	14 HPS plants with 1,380 MW capacity are installed in Switzerland. The estimated energy storage is 369 GWh. In this research, Switzerland's HPS sector is studied in conjunction with the German electricity system, by taking into account the intermittent operation of RES (solar, wind and mix). The overall system is assessed vs the loss of load probability (LOLP) and the energy shortages.	The German electricity system becomes more stable when the storage plants are added to the German electricity system. The loss of load probability is considerably reduced for all three RES scenarios (solar, wind and mix). Moreover, there is a drop in energy shortages and energy surpluses. Energy surpluses are stored in the HPS and this allows for a production increase, that is, lowering the shortages.	(van Meerwijk *et al.* 2016)
Adele project, Sachsen-Anhalt, Germany	260 MW plant with CAES storage	The discharge time of CAES is equal to five hours and the efficiency is equal to 70%.	(Molina 2017; DoE 2016)
Joint European Torus (JET) Fusion FES facility in Culham, Oxfordshire	Flywheel energy storage system of 400 MW capacity	Discharge time of around 30 seconds at rated power, resulting in 12 000 MJ stored energy. The plant is used to provide power balance services to the Culham Centre for Fusion Energy (CCFE).	(Molina 2017; DoE 2016)

Table 8.2. *Case studies of mechanical energy storage*

8.1.3. *Thermal energy storage in smart grids*

Thermal energy storage is divided into sensible, latent, and thermochemical energy storage, as depicted in Figure 8.1. An overview of thermal energy storage systems (TES) is provided by Alva *et al.* (2018).

Sensible heat energy is stored by changing the temperature of the storage materials. The amount of stored energy depends on the medium's density, specific heat, volume, and temperature variation (Li 2016). Sensible storage can be stored in solids (stones, salt), liquids (molten salts, water, thermal oil) and liquids with a solid filler (water with pebbles, molten salt with oil, and so on) (Pfleger *et al.* 2015). Water is one of the commonly used materials for heat storage in residential applications, while water tanks and aquifer storage systems are the most commonly used water-based storage systems. Other materials used for sensible energy storage are depicted in Figure 8.2.

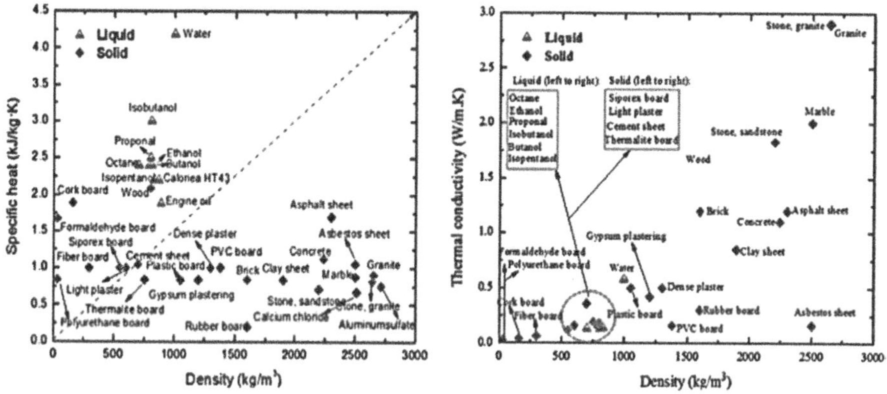

Figure 8.2. *Thermal properties of sensible heat materials (Li 2016; Kalogirou 2004)*

Concerning sensible thermal storage systems, heat pumps can be an indicative example seen as part of the demand side that can be actively managed to support the realization of a smart grid (Fischer and Madani 2017). When heat pumps are utilized in conjunction with thermal storage, they offer the possibility to decouple electricity consumption from heat demand. This is very useful in increasing the performance and renewables penetration in a smart grid.

Molten salts are the most commonly used storage material in concentrated solar power (CSP) applications due to the thermal properties and cost (see Table 8.3) (González-Roubaud *et al.* 2017). An example of molten salts thermal storage coupled with CSP is analyzed in section 8.2.2.

	Line focus technologies		Point focus technologies		
Design	**Parabolic trough**	**Linear Fresnel**	**Solar tower**	**Beam down**	**Dish–stirling engine**
Technology maturity	Most mature	Few installations	Commercial deployments	Early development	Few installations
Preferred scale	Large	Medium-large	Large	Large	Small-medium
Capital Cost	Moderate	Low	High	Moderate to high-low storage cost	High (low per unit)
Operating cost	High	Low	High	Similar to solar tower	High (one engine per dish)
Annual solar to Net electricity conversion efficiency (if connected to a power block)	~15%	~11%	~17%	~15%–19%	22%
Thermal storage	Feasible	Feasible	Feasible and more efficient due to higher temperature	Feasible with very little energy cost	Not currently possible
Characteristics	– Significant construction and operational experience – High radiative and convective energy losses	– Low cost due to fewer moving parts – Lower efficiency – High temperature heat transfer fluid possible	– High cost due to expensive heliostat field – High temperature heat transfer fluid possible – High efficiency	-Lower storage cost	– High engine efficiency – High cost due to expensive engines (one for each dish)

Table 8.3. *Advantages and disadvantages of concentrated solar power (CSP) technologies*

Latent thermal energy is stored by changing the phase of the storage materials. The latent storage applications usually exploit the phase change between solid and liquid phases to store/release thermal energy, due to the higher heat storage capacity per volume and low volume difference between these two phases (Moslehi and Kumar 2010). The main advantage of latent versus sensible storage is the higher storage capacity per unit of mass (and volume), resulting in lower space requirements (Flores *et al.* 2017). Also, as phase change occurs, there is only a small temperature difference during the charging or discharging process.

Phase change materials (PCMs) have become more and more attractive over the last few years for space heating and cooling in buildings, solar applications, off-peak energy storage, and heat exchanger improvements (Rousse *et al.* 2009; Pisello *et al.* 2017; Guarino *et al.* 2017). PCMs are substances that can exist in solid, liquid or gaseous states depending on the temperature and pressure of the storage conditions. An example of using PCMs at the district level for low temperature heating is included in Flores *et al.* (2017), showing that PCMs have better performance in district heating applications than sensible storage systems, that is, water purifiers. A significant drawback is the increased cost of the PCMs, especially in specific temperatures.

The thermochemical energy storage has received considerable attention in the last decade (Mette *et al.* 2012). Long-term thermal energy storage, for example, seasonal storage for solar thermal applications can increase the fraction of the solar energy utilization factor from 20%–30% to 50%, or even 100%. This can be achieved by thermochemical energy storage, where the heat is stored in chemical bonds of special materials and the charging/discharging process is an endothermic or exothermic reaction (Pardo *et al.* 2014).

Some examples of integrating thermal storage in smart grids/microgrids are presented here (Gibb *et al.* 2018):

– A low temperature solar power plant with temperatures between 80 and 130°C is coupled with a Sofia-Antipolis smart grid designed in the framework of the PREMIO project. This solar plant is also connected to a sensible heat water storage device of 150 kWh_{th} and a 5 kW_e organic Rankine cycle (ORC) turbine. The water storage device is utilized to provide energy, even when there is no energy production from the solar plant (Haillot *et al.* 2011).

– An underground aquifer thermal energy storage system (ATES) that can cover the energy demand of group of buildings is analyzed in Rostampour and Keviczky (2017). This storage system can be applied as a seasonal storage system in moderate climatic conditions where there is shortage of heat during winter and an excess during summer. The ATES can therefore be considered as a heat source or sink, or a storage for thermal energy depending on the season and the needs.

– Sensible thermal storage can be provided by the building envelope in the form of thermal mass and/or insulation. Typical new residential buildings in Belgium equipped with air-to-water heat pumps that supply either radiators or a floor heating system are considered in Masy *et al.* (2015). The analysis is based on the energy cost reduced, as well as on the loads' flexibility. The flexibility of the loads is quantified by the amount of load volumes shifted. By formulating the buildings in a smart grid typology and applying smart grid control strategies, a reduction of 13% of energy costs by the consumer is reached. The flexibility is increased by 3% to 14% with the same thermal comfort.

– An activated building wall, PCM, hot water tank and thermochemical material (TCM) storage are investigated to optimize the scheduling of thermal storage and minimize the energy cost of an office building connected to a smart grid. The cost reduction is estimated to be equal to 12.5% (Finck *et al.* 2015).

– The grid of the University of Genoa is used to optimize the combination of renewables and thermal storage, in order to cover the thermal loads on yearly basis by minimizing capital and variable costs. The grid consists of a 100 kWe micro gas turbine, a 20 kWe internal combustion engine for electricity and hot water and a 100 kW_{th} adsorption chiller for cooling. The grid is also equipped with a small size wind turbine and photovoltaic solar panels. The optimized sizing of renewables indicates a 17 kW PV installation and a 3 kW wind turbine. The sizing of renewables indicate that the optimum size of thermal storage is equal to 30 m^3 (Rivarolo *et al.* 2013).

– Ice storage is used to provide cold water to the business district of Paris La Defense through a distribution network linked with a central refrigeration system of three units, with a unit capacity of 8 MW. The storage is performed daily with a charge temperature of −7°C and discharge temperature of 4.5°C. The ice storage capacity is equal to 240 MW with an efficiency of 95% (Gibb *et al.* 2018).

8.2. Energy storage and smart grids: case studies

8.2.1. *Case study 1: the Leaf Community smart grid energy storage system*

This case study involves the micro-grid of the Leaf Community, in Angeli di Rosora, Italy (Figure 8.3). The energy production sources connected to the grid are a micro-hydropower plant of 48 kWp, four rooftop PV installations of total 488.6 kWp, a vertical PV of 10.7 kWp and a dual axis Solar Tracker of 18 kWp. The Leaf Community consists of five buildings, which are interconnected with the energy production technologies mentioned above. Moreover, EVs are used for the transportation of the Leaf Community employees. The Leaf Community is equipped with:

– an electrical storage system of 224 kWh;

– a thermal storage system with a heat capacity of 1883.7 MJ/K.

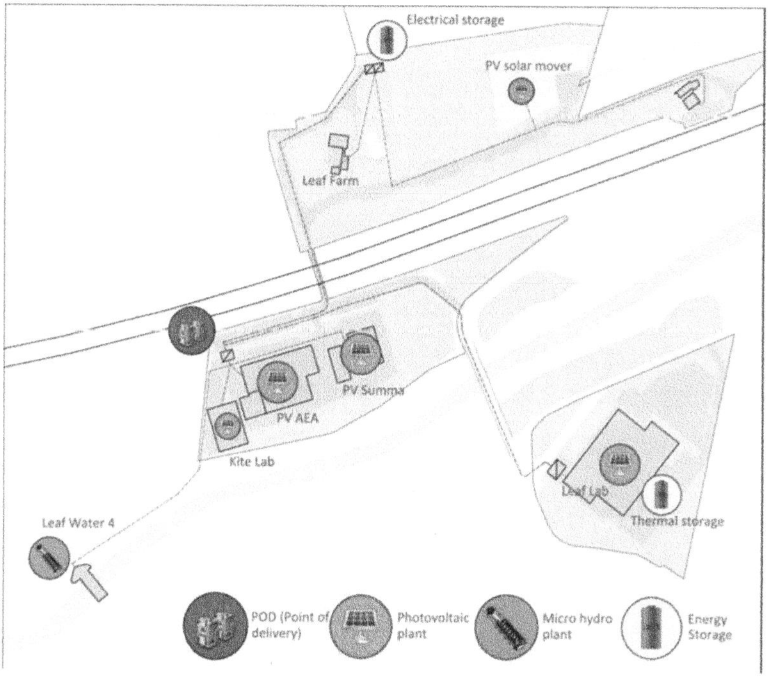

Figure 8.3. *The Leaf Community smart grid*

All of the previously mentioned power loads, renewables and storage components are in parallel with one single point of delivery (POD). All smart grid components are monitored and controlled via a web-based monitoring platform.

8.2.1.1. *Leaf Community web-based monitoring and control platform*

The Leaf Community's smart grid is controlled by centralized software called the energy management system (EMS). The EMS is organized into three levels, as depicted in Figure 8.4. This architecture allows a reliable, efficient, and flexible control of a heterogeneous cyber-physical system.

Leaf Lab HVAC Units – Model type	Heating power (kW)	Heating COP	Cooling power (kW)	Cooling EER
NECS-WN/S 0412	130	4.8	159.2	6.89
NECS-WN/S 0904	286.9	4.19	239.3	6.19
NECS-WN/S 0904	286.9	4.19	239.3	6.19

Table 8.4. *Leaf Lab heating, ventilation, and air conditioning system (HVAC) heat pumps technical characteristics*

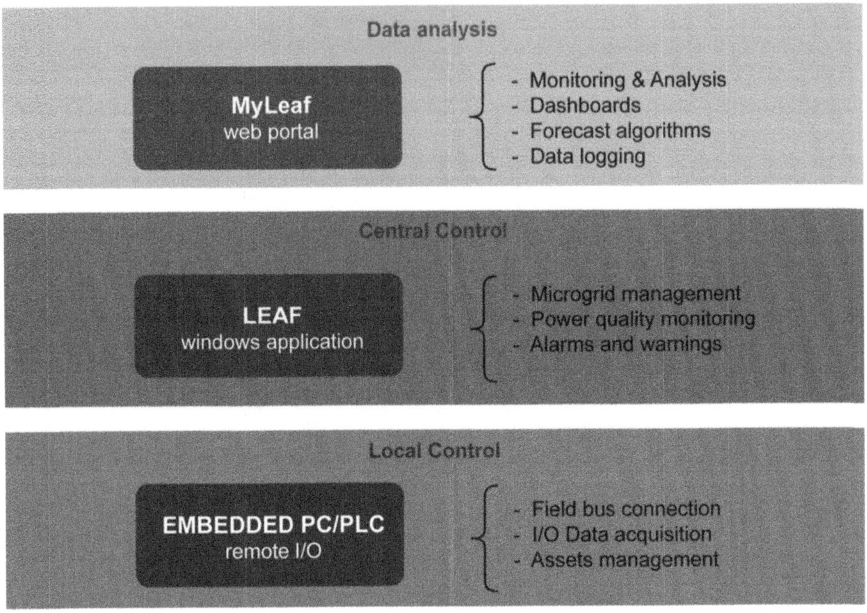

Figure 8.4. *System architecture*

8.2.1.2. *Local control*

Each microgrid element is equipped with a local control composed by a PLC and an embedded PC. The PLC reads the data and executes the real time operational functions necessary for the proper functioning of the plant. The PC allows the plant control with a local human machine interface (HMI). The communication between central and local control is realized by REST Web Services (Giessler *et al.* 2016).

8.2.1.3. *Central control*

The facility operator uses centralized software to monitor the smart grid and execute several control logics, in order to optimize the energy fluxes. The central control reads all real time data from each local control PC. Information is organized in logical representations of a physical asset (PV plant, Micro-Hydropower plant, Electrical Storage) characterized by properties and behaviors.

The software combines behaviors, data from the grid and forecast data, in order to implement the control logics and, moreover, sends commands to distributed energy resources (DER). Examples of control for the storage are:

– peak shaving: cuts the power consumption peaks;

– self-consumption: maximizes the amount of energy self-consumed;

– max export/max import: the target is maximizing the energy produced/ consumed by Leaf Community.

8.2.1.4. *Data analysis*

The central control sends data to a central platform. This platform stores long term data and provides validation and reconstruction algorithms. The facility operator can visualize historical and real-time data via a web browser in order to define KPIs and analyze the grid's performance.

8.2.1.5. *The TES of the Leaf Community*

The TES is a water tank with dimensions 12.3 m × 11 m × 3.4 m and a capacity of 400 m^3 (Figure 8.5).

The water tank is buried and insulated with 16 cm of extruded polystyrene and is used to store the sensible heat for the energy demand of the connected buildings.

TES was initially used to implement load shifting, by charging the TES during the night, when the energy cost is 10% less than during the day, and discharging it during the day. This usage was not considered efficient due to the fact that the TES losses are higher than the energy cost reduction during the night. The thermal energy losses from the water tank are calculated to be equal to 12%, due to conduction. An extra 25% is lost by the GSHPs due to the fact that in order to charge and discharge the thermal storage, it is necessary to increase and decrease the GSHP output temperature set-point during winter and summer respectively. Therefore, the surplus power produced by renewables is considered the most cost-effective solution. The energy buying price for RES is 70% higher than the selling price to encourage self-consumption strategies. The Leaf Microgrid only has a surplus production of energy during weekends and holidays. A performance analysis of the TES is included below.

8.2.1.6. *Electrical energy storage in Leaf Community*

The EES consists of Lithium 220 kW/kWh (Figure 8.6). The specific energy storage system is used for the optimization of the energy flows and peak shaving. Its optimized operation is described in Kolokotsa *et al.* (2015). When the charging-discharging of the batteries is optimized based on their swing range and life cost, while energy demand and production for the Leaf Community is predicted using artificial neural networks, a 6% reduction of the energy cost is achieved.

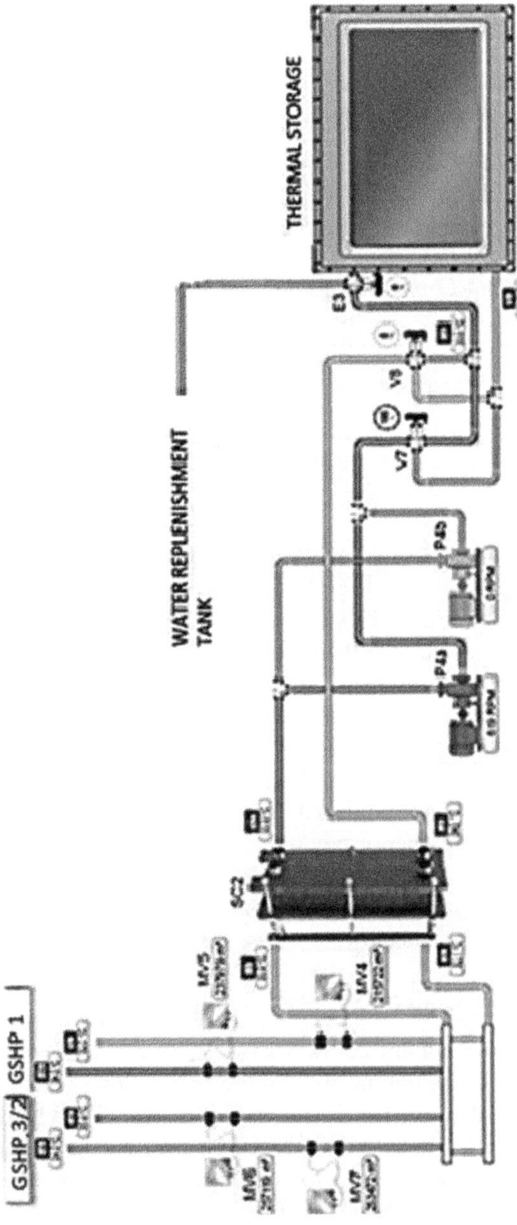

Figure 8.5. *Thermal storage – ground source heat pumps (GSHP) connection schema*

Figure 8.6. *The batteries storage system in the Leaf Community*

8.2.1.7. *Performance analysis of energy storage for the Leaf Community microgrid*

8.2.1.7.1. Performance analysis of the TES

The TES is mainly exploited to reduce the peak power of the Leaf Community's interconnected buildings.

One of the buildings' energy demands that is covered by the TES is the Leaf Lab. The Leaf Lab is a 6,000 m^2 building equipped with ground source heat pumps (GSHPs), chilled beams, and fan coils. The Leaf Lab heating, ventilation, and air conditioning system's (HVAC) specifications are tabulated in Table 8.4.

The excess energy produced by the Leaf Community's energy production sources during weekends, holidays, and so on is used to operate the heat pumps and store heating or cooling energy in the thermal tank.

The stored energy is then used to optimize the HVAC efficiency and reduce peak demand during working hours. The overall operation is described below:

1) The thermal storage is charged during the weekends using the excess energy produced by the Leaf Community.

2) This excess production is used to operate GSHP2 and GSHP3. The charging process begins when there is a PV power production that exceeds 60 kW. This is the threshold for the GSHP3 activation.

3) After the activation of the GSHP3 and if there is an excess of 50 kW, GSHP2 is activated.

The activation of the heat pumps for charging the thermal storage is only allowed during weekends, from 8:00 to 16:00 in winter and 7:00 to 18:00 in summer. The pumps are switched off at the end of each schedule or if the PV production is significantly reduced over a specific time period. In case the PV power is instantly reduced, power is withdrawn from the grid in order to keep the heat pumps under operable mode.

For the deactivation of the heat pumps, if the power from the grid is greater than 130 kW, GSHP3 is switched off. Then, if the energy withdrawn from the utility grid exceeds 90 kW, the GSHP2 is switched off.

The TES charging status and temperature is depicted in Figure 8.7.

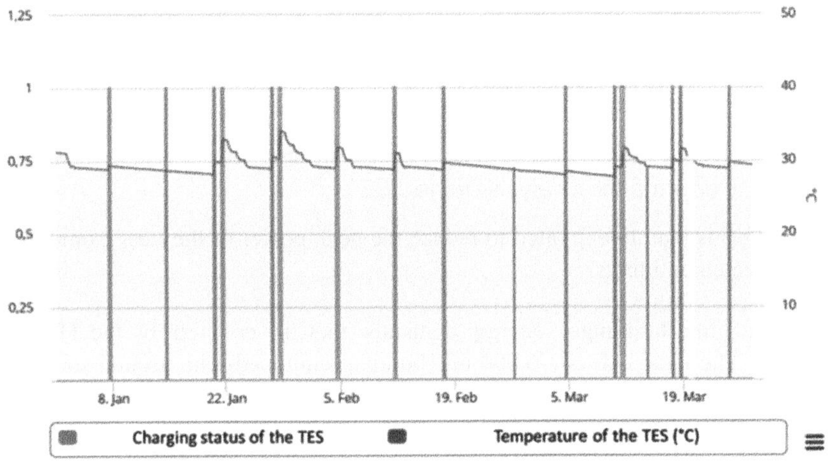

Figure 8.7. *TES charging during weekends and temperature of thermal energy storage system (TES)*

TES performs better during spring and autumn because the outdoor air temperature is close to the indoor temperature set-point and the HVAC system is not stressed. The thermal energy stored during the weekend can support the HVAC needs for the whole week. Figure 8.8 depicts typical TES power exchange with the grid during a spring week.

Positive values represent the charge of storage on Saturday and Sunday, while negative values represent the discharge during the week days. During summer and winter, TES is sufficient for one or two days depending on the outdoor weather conditions. TES covers almost 25% of the Leaf Lab building's energy demand.

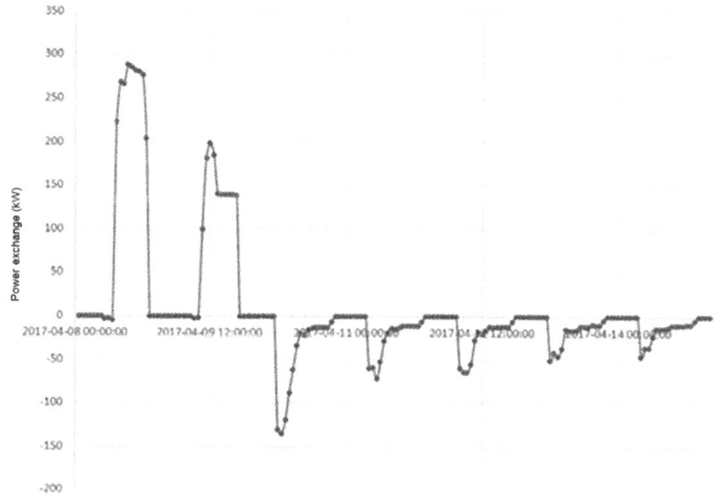

Figure 8.8. *Thermal power exchange with the leaf microgrid*

8.2.1.7.2. Performance analysis of the BES

The BES is either used for peak shaving or self-consumption. Peak shaving depends on the maximum electricity power peak measured by the Leaf Microgrid POD and is almost 610 kW. Since the nominal power of EES is 220 kW, the EES is activated when the electricity power is higher than 450 kW.

The peak shaving mechanism operates in the following way:

– if the microgrid consumes more than 450 kW, the electric energy storage is activated in order to keep the energy consumption under the specific threshold;

– if the microgrid consumes less than 150 kW, then the electric storage is recharged.

The peak shaving operation is depicted in Figure 8.9. Each time the POD power exceeds 450 kW the battery is discharged to support the extra power requirements. Also, during the night, when the microgrid's energy demand is reduced, the batteries are charged.

Sometimes the EES energy capacity is not enough to guarantee the performance over time. In Figure 8.10, the POD power, SOC of battery and power of battery are depicted. When the SOC is 30% EES is not activated, which is the case on 22 November after 16:00. This means that the EES nominal power should be increased to cover peak shaving demands more effectively.

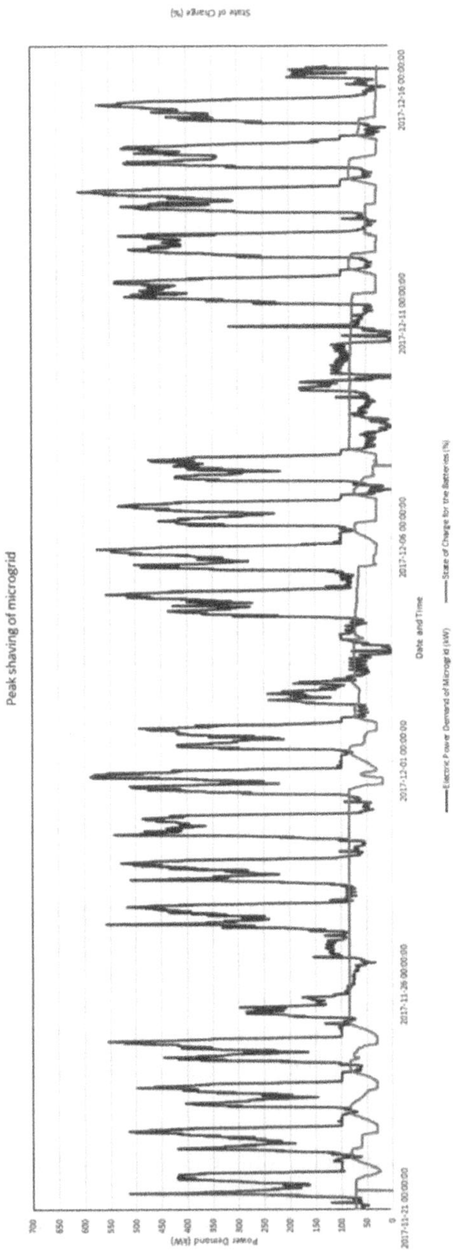

Figure 8.9. *Peak shaving of microgrid using BES*

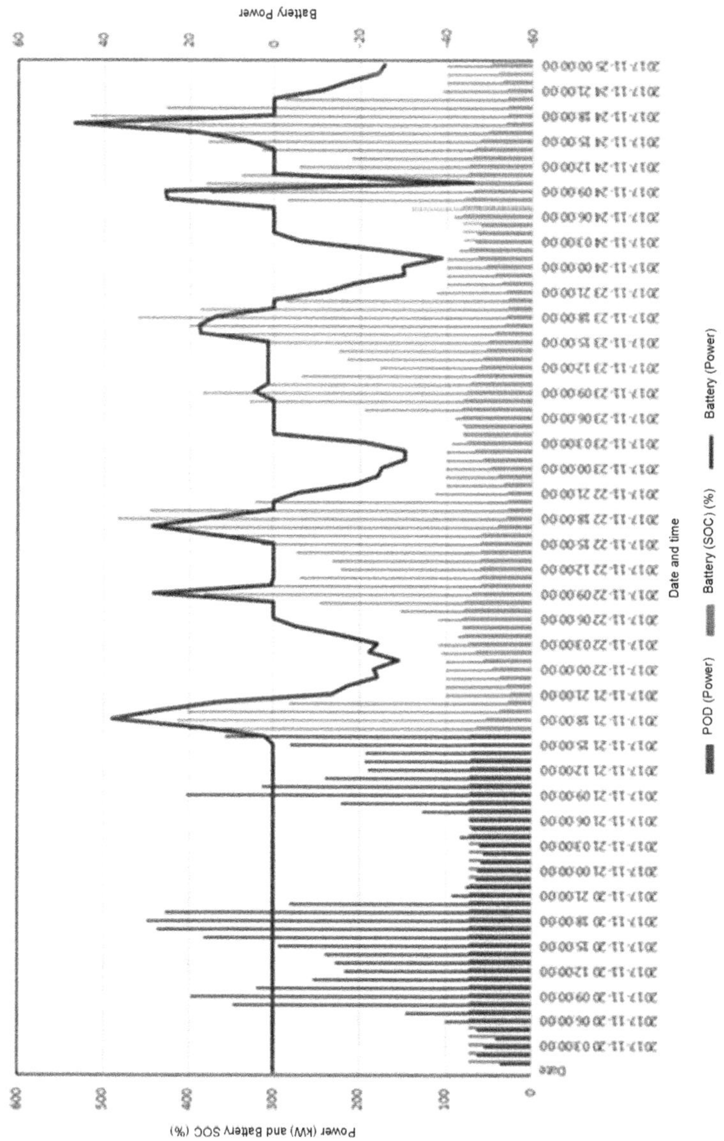

Figure 8.10. *Peak shaving strategy*

The self-consumption strategy aims to maximize the self-consumed energy. The Leaf Microgrid only discharges power to the grid during the weekend. A branch of the 116.65 kW photovoltaic plant and a little building called the Ramo Leaf Farm with 30 kW peak consumption is formulated. The EMS is configured to perform a guaranteed daily energy exchange between the BES and the branch. This means that the energy stored in BES is self-consumed by the branch. The self-consumption strategy showed that Ramo Leaf Farm consumption in 2018 reduced by 33% compared to 2017, due to the maximization of the PV plant energy self-consumption, as depicted in Figure 8.11.

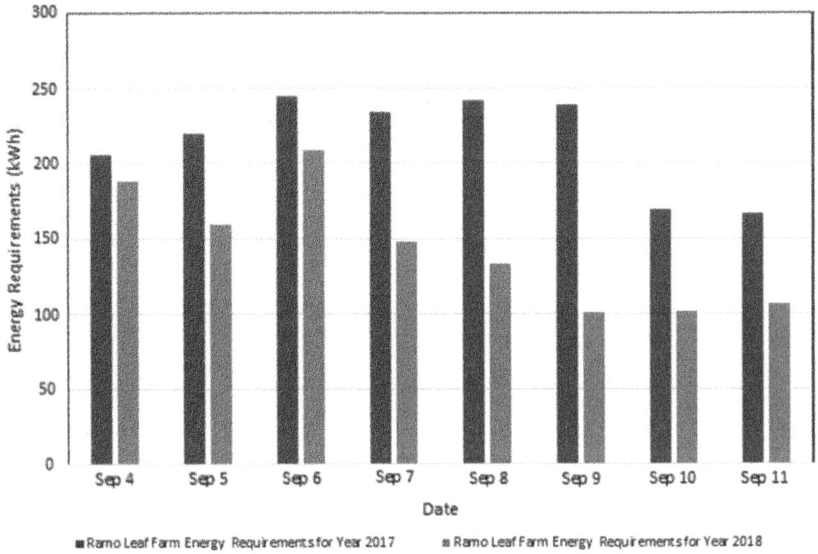

Figure 8.11. *Performance of self-consumption*

8.2.2. *Case study 2: energy storage of CSP and integration with smart grids*

8.2.2.1. *Technical description of a polygenerative solar plant*

The technology described in this section is developed and tested by IDEA Srl. The solar plant consists of a compact collector for multi-generative applications, which is based on a Linear Fresnel concentration system. The solar rays are reflected to the same target by means of appropriate mirrors that constantly track the solar position during the day. The collector has been optimized to reduce the space required for its installation thanks to a compact and lightweight design and a reduced focal length.

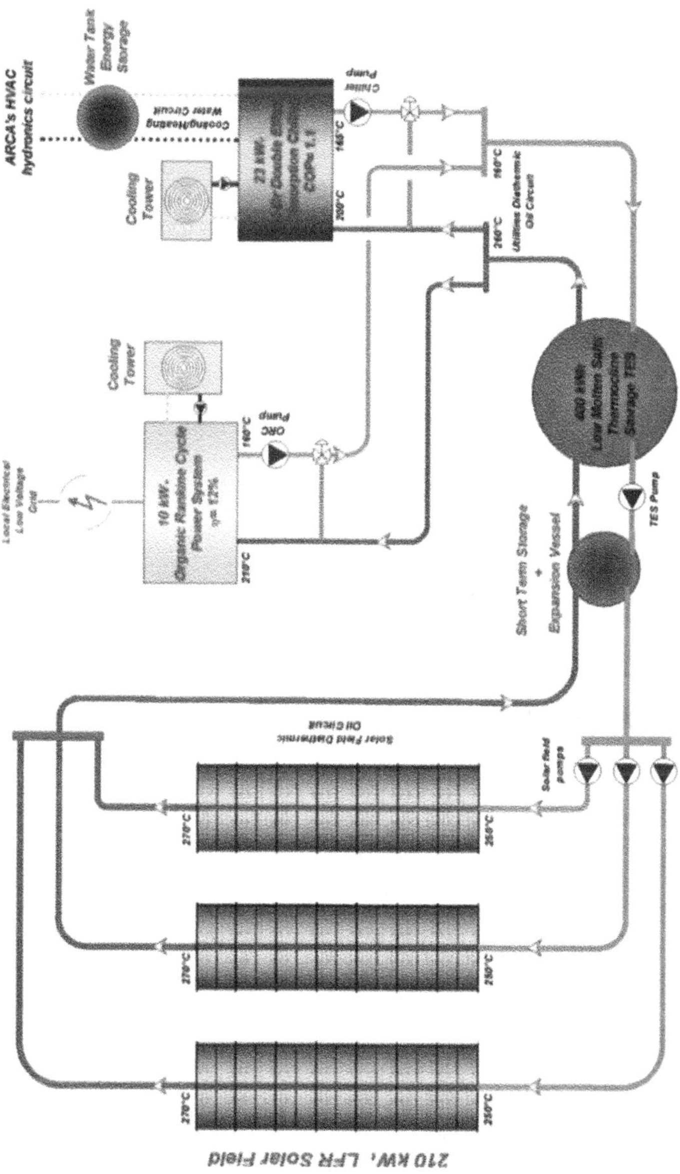

Figure 8.12. *The poly-generative system outline*

The solar collector is able to concentrate the solar radiation into a receiver tube in which a high temperature HTF is heated. The thermal energy is collected at about 270°C and is stored inside a storage tank. This stored energy can be utilized either by absorption chillers for cooling or directly by heating infrastructure.

Additionally, with the concentrated solar energy, solar heat is converted into power by running thermodynamic cycles. Thus, the system can produce electricity through an ORC unit. The schematic of the polygenerative plant is depicted in Figure 8.12.

8.2.2.2. *The polygenerative solar plant in a university campus*

The polygenerative solar plant described in section 3.3 has been installed in the campus of the University of Palermo, Italy as part of the University's Solar Living Lab at the ARCA business incubator (STS–Med 2018). It consists of three strings with seven modules each, for a total collecting surface of about 480 m^2, as depicted in Figure 8.13 (Kolokotsa *et al.* 2016b; Gobakis *et al.* 2017). All parts of the installation are described in the following sections.

The design parameters of the solar plant are tabulated in Table 8.5. The monthly solar energy produced by the LFR is tabulated in Table 8.6.

The primary optic of each of the 21 modules is composed of 18 curved mirrors, 32 cm wide and 400 cm long, with a total reflectance of about 95%. Reflective panes are fixed on a steel structure that is connected to the tracking system. Mirrors with different curvatures are installed from the center to the extremity of the module, in order to modulate the focal length according to the respective position of the absorber tube placed at about 3.5 m above the module plane.

The secondary optic consists of a metal mirror with a solar reflectance of about 84%. It has been appropriately designed with a compound parabolic collector (CPC) configuration, in order to refocus the imperfectly collimated solar rays on the absorber tubes. The absorbers are steel pipes with cermet coating, vacuum glazed. The peak thermal power of the solar field is 220 kWth with an estimated annual thermal energy production of 295 MWh_{th}. The HTF is a synthetic oil with a specific heat coefficient of 2 kJ/(kgK) and a maximum outlet temperature of 310°C (Fedak *et al.* 2017).

When selecting oil as a transfer fluid, the main limiting factor, to be taken into account for the maintenance of stability, is the maximum temperature of the oil, as above this temperature, decomposition and rapid chemical degradation occurs.

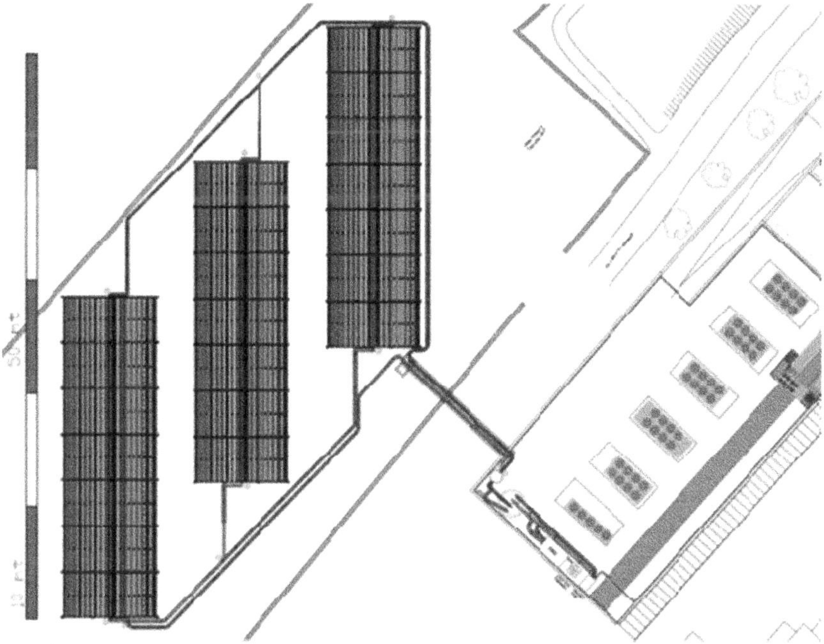

Figure 8.13. *The layout of the installation site at IDEA srl. (Long 38.10o Lat 13.34o). The three strings cover an installation of total 480 m^2*

The advantages of hot oil utilization in solar energy generating systems (SEGS) compared to water utilization in direct steam generation (DSG) are (Pye 2008; Zarza *et al.* 1997) relevant:

– With the use of synthetic oil, possible instability due to two-phase flow of water/steam in the collector is avoided. Moreover, possible process instability due to large changes in fluid volume during boiling is minimized. Furthermore, in DSGs, high temperature wet and/or dry conditions require higher-grade materials, and if superheated steam is generated in the collector, there is a not negligible risk of thermal stress.

Nevertheless, the oil as HTF has the following disadvantages:

– there is an increased risk of fire and pollution;

– the upper temperature limit of ~300 to 400°C required for the use of synthetic oil should be continuously monitored;

– there is a need to replace almost 5% of the oil each year;

– anti-freezing is also needed when the temperature is below 14°C.

The upper temperature limit imposed by the stability range of the oil directly implies an upper limit for the inlet temperatures in the ORC.

8.2.2.3. *Short-term thermal energy storage – the buffer tank*

When the heated oil exits the absorber tube, the HTF enters a short-term buffer storage of 800 L (see Figure 8.12) as the effective specific heat of the HTF increases.

Outer absorber tube diameter (m)	0.07
External diameter of glass cover (m)	0.125
Receiver coating absorbance (%)	0.958
Glass cover transmittance (%)	0.964
Glass mirror width (m)	0.32
Primary optic reflectance	0.958
Secondary optic reflectance	0.84
Module length (m)	4
Number of mirrors per row	18
Module focal length (m)	3.5
Total number of modules	21

Table 8.5. *Fresnel parameters used for the efficiency calculation*

This first buffer stabilizes the exchange processes within the LFR. The integration of a buffer storage capacity simplifies the control of the power plant and allows an extended reaction time for backup systems, which are intended to compensate longer periods of reduced insolation (Steinmann and Eck 2006).

Month	Solar energy reaching the mirror surface (kWh)	LFR solar thermal energy gain (kWh)	Monthly average efficiency (%)	Thermal peak power (KW)
January	40,527	3,215	8%	63.56
February	42,064	6,040	14%	87.85
March	60,381	16,429	27%	153.64
April	73,345	27,106	37%	189.14
May	104,554	45,443	43%	199.12
June	112,330	49,264	44%	220.60
July	133,208	58,950	44%	221.39
August	106,128	44,547	42%	191.31

September	77.963	24.436	31%	169.66
October	64.843	12.225	19%	127.89
November	55.120	5.375	10%	70.57
December	43.923	2.881	7%	47.61
Total		295.911		

Table 8.6. *Monthly solar thermal energy, efficiency, and thermal power of IDEA's linear Fresnel reflectors (LFR) system in Palermo*

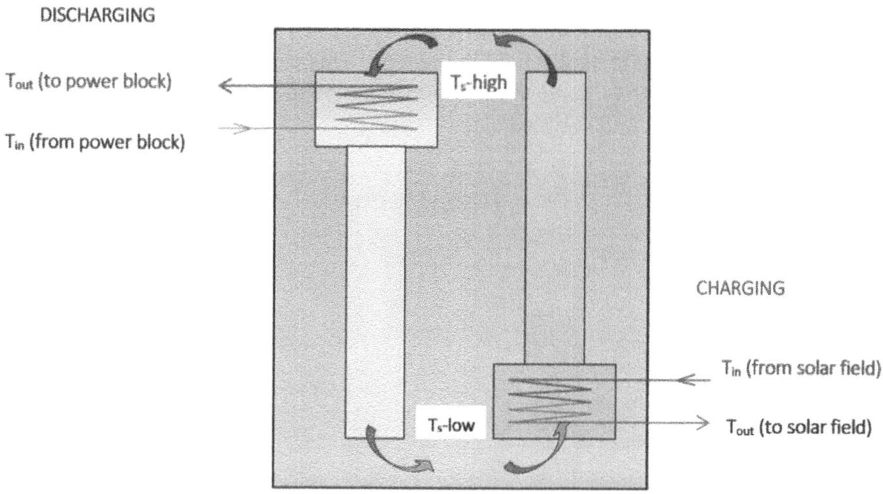

Figure 8.14. *Schematic illustrating principle of operation of the thermocline thermal storage installed in Palermo*

This short-term storage is important for damping fluctuations in power output associated with short-term disturbances, such as passing clouds. The short-term thermal storage mechanism uses a pressurized vessel and a simple, inexpensive storage medium. In the specific LFR system, pressurized N_2 is used at 3 bars. Storage is limited to small capacities in the order of some hours.

8.2.2.4. *The molten salts thermocline storage*

The heated oil from the receiver tube, after exiting the first short-term buffer, is diverted to a heat exchanger to heat a TES fluid, typically a Molten Salt (MS) mixture. MS-TES systems are generally built with two separate tanks (cold tank at 290°C and hot tank at 380°C when oil is used as heat transfer fluid).

Unfortunately, molten-salts freeze at relatively high temperatures 120°C to 220°C. This means that special care must be taken to ensure that the salt does not freeze in the solar field piping during the night. Newer linear Fresnel designs may allow the use of higher-temperature molten salts.

An innovative and simplified solution has been adopted in the polygenerative solar plant installed in Palermo. A particular ternary mixture of nitrate salts with a low melting temperature is used in a single stainless steel thermal tank with a diameter of 1.8 m and a height of 3.5 m. It contains 8 m^3 of salt, as well as equipment such as sensors, electric resistances and heat exchangers.

An eutectic mixture of $CaCO_3/NaNO_3/KNO_3$ (42/15/42% w) is heated by the heat transfer fluid from the solar field in the tank. This particular salts' mixture, with a melting point of 130°C (and a maximum temperature of about 450°C), allows both an easy operation and maintenance.

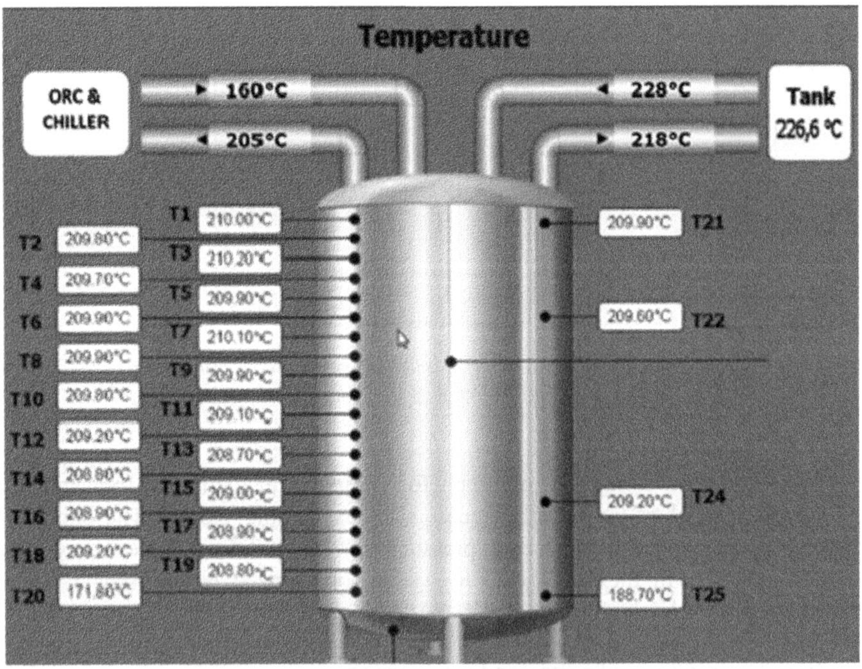

Figure 8.15. *Distribution of temperatures profile in a charged tank installed in Palermo*

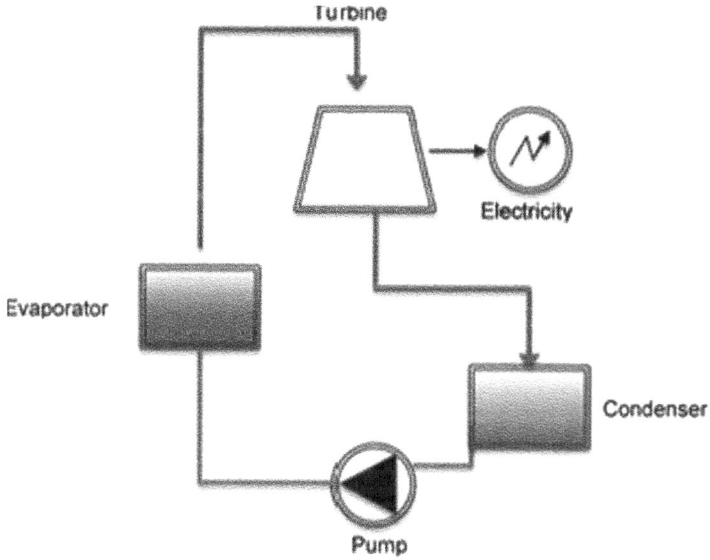

Figure 8.16. *The organic Rankine cycle (ORC) generator*

The natural stratification of the heat into the thermal storage (thermocline effect) allows the charging and discharging process in the same storage using the natural temperature gradient difference. Two different heat exchangers are installed, respectively, the discharging heat exchanger with a power of 150 kW at the top of the tank and the charging heat exchanger with a power of 200 kW at the bottom of the tank (Figure 8.14).

During the day, the hot oil coming from the solar field heats the molten salts transferring the thermal energy gained by the collectors to the storage.

The molten salts thermal storage is equipped with two heat exchangers specifically designed with coils in AISI 316 stainless steel. The exchangers are respectively connected to the oil buffer tank, for the TES charging, and to the equipment (absorption chiller and ORC turbine) for the TES discharging. Both of the heat exchangers are helicoidally shaped and designed for an optimized heat exchange process between the oil flow and the molten salt mixture zone, either during the charging or the discharging phase. Two vertical channels of 7.62 cm (3 in.) diameter, directly connected to each heat exchanger, allow the salts to transport to the top and to the bottom, during the charging and the discharging phases, respectively.

A total storage of 400kWh$_{th}$ is achieved with 16 tons of hydrated salts, capable of driving a small double effect chiller (23 kW$_{th}$ – COP 1) and a small ORC (10 kW$_e$ with a net efficiency of 10%) described in the next sections.

The thermal distribution into the tank is monitored by a group of 20 sensors (PT100) immersed with a mutual interdistance of 15 cm in the salt mixture. An extra set of four temperature sensors measure the thermal profile close to the charging heat exchanger, in order to evaluate any thermal difference in their layer. Figure 8.15 shows the distribution of the temperature into the thermal storage, recorded on a typical operating day during the months of June and July 2018.

The operational tests are conducted with the contribution of solar energy produced by LFR collectors at different inner temperature profiles. At this scale, tests have also confirmed the effective production of a thermocline effect during the charging phase, with a natural stratification of the temperatures from the top to the bottom and a constant thermal power exchange during the whole process. The exchange of thermal power is calculated by the inlet and outlet oil temperature to the TES and the values of specific heat and density. The natural stratification of the MS bulk temperature is recorded by T19–T1 sensors, while the heat stored in the upper volume influences the sensor T20 less, which is placed close to the shell in the bottom zone.

Thermal energy stored in the tank is used to provide thermal energy after sunset or during cloudy periods. When the storage is fully charged, it is by-passed and the solar field drives the equipment for cooling or power generation directly.

The thermocline single tank approach could overcome the traditional “two molten salt thermal tanks” design, usually adopted in the CSP solar fields.

New applications of binary and ternary mixtures, with a lower melting point, have been discussed from a technical and economic perspective. The target of these applications is to compare the efficiency of single tank thermocline molten salts storage and commercial “two tank” configurations (Angelini *et al.* 2014).

In such studies, thermocline storages systems with nitrates salts, used in the range of 300° to 400°C, with quartzite fillers, have been considered economically advantaged due to the reduced costs of salts, materials, and equipment (Pacheco *et al.* 2002). Nevertheless, some flexibility in the design is required for the adoption of a single tank approach, as the two tank configurations represents the most suitable solution to keep the plant working close to the design point (Kolb 2011).

The innovative MS-TES installed in Palermo is designed for an operative range of 200°C to 300°C, and is able to drive equipment operating in a flexible range of 160°C to 200°C. The aim is to investigate the application of medium temperature thermocline storage in civil and industrial solar energy plants. The benefits of a new generation of cheaper and simple thermal batteries could also be valuable in recovering wasted thermal energy from many industrial processes.

8.2.2.5. *The absorption chiller*

An absorption chiller is integrated in the solar field in order to serve the cooling loads of the offices during the summer season. Thermally driven chillers, using solar energy, are a good alternative compared to conventional electricity driven chillers to meet the growing air-conditioning demand and cut electrical peak loads during the summer. Also, their use allows primary energy to be saved and greenhouse gas emissions associated with the electricity generation from fossil fuels to be reduced (Montero-Izquierdo *et al.* 2011).

Solar thermal cooling systems have been widely used, particularly absorption cooling systems. Solar cooling using single-effect water-LiBr has been studied since the mid-1970s and until now, many studies have been performed.

In literature, applications of various configurations have been reported, such as (Montero-Izquierdo *et al.* 2011):

– flat plate collectors and single-effect chillers;

– evacuated tube collectors and single-effect chillers;

– compound parabolic collectors (CPC) and single-effect chillers;

– parabolic trough collectors (PTC) and single-effect chillers;

– parabolic trough collectors and double effect chillers and systems;

– linear Fresnel collectors (LFC) and double-effect chillers.

While single-effect absorption chillers work at relatively low driving temperatures (between 80°C and 95°C) produced by efficient flat plates or vacuum tube collectors, and delivering low COPs (between 0.55 and 0.75), double-effect absorption chillers demand a higher driving temperature (up to 160°C), which results in higher overall efficiency (COP = 1) (Montagnino 2017).

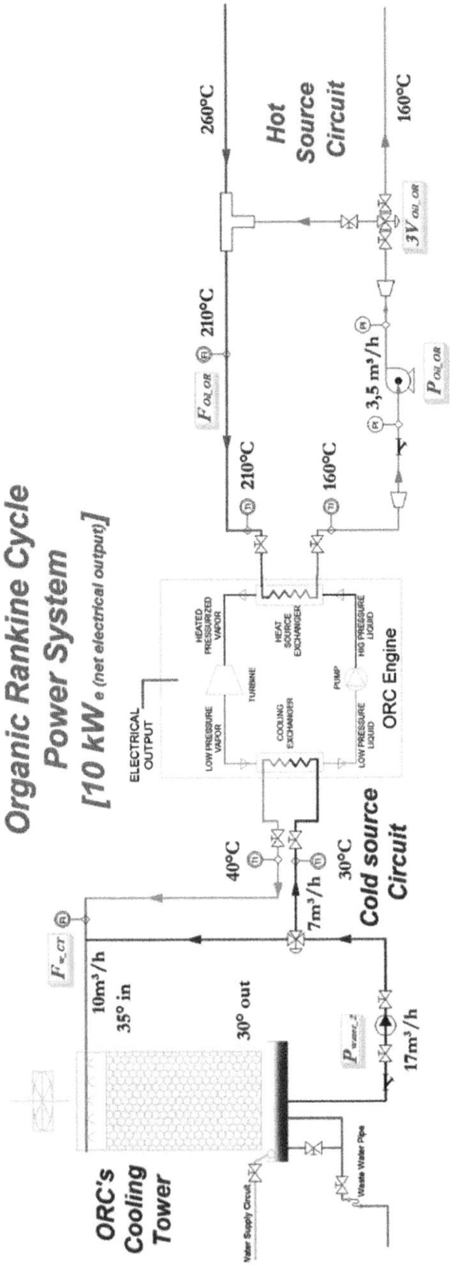

Figure 8.17. *Organic Rankine cycle (ORC) section of the pilot plant in Palermo*

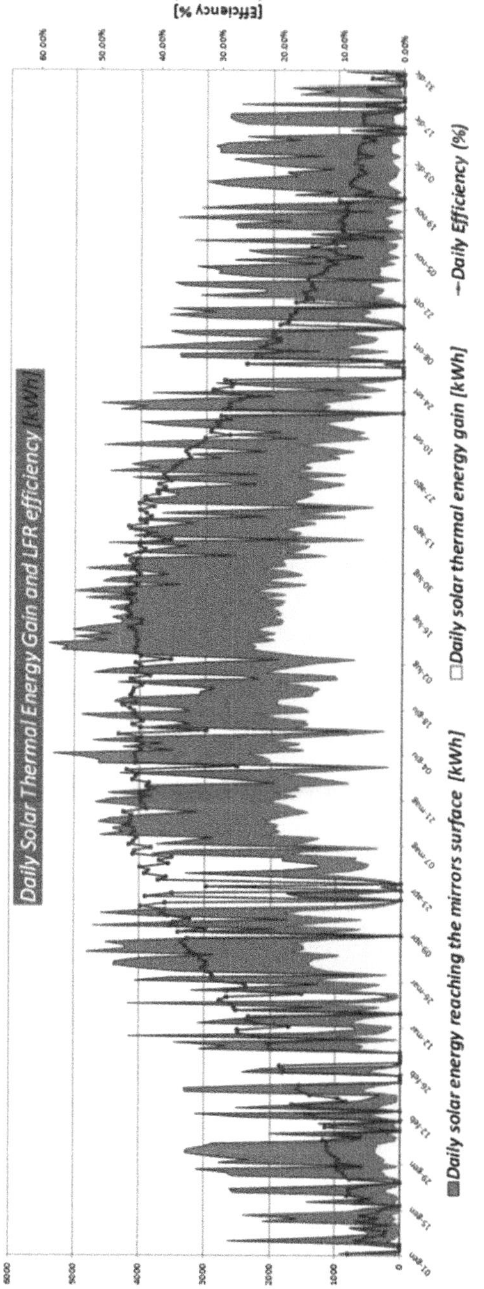

Figure 8.18. *Steady-state calculation of daily solar thermal energy gain and linear Fresnel reflectors (LFR) efficiency*

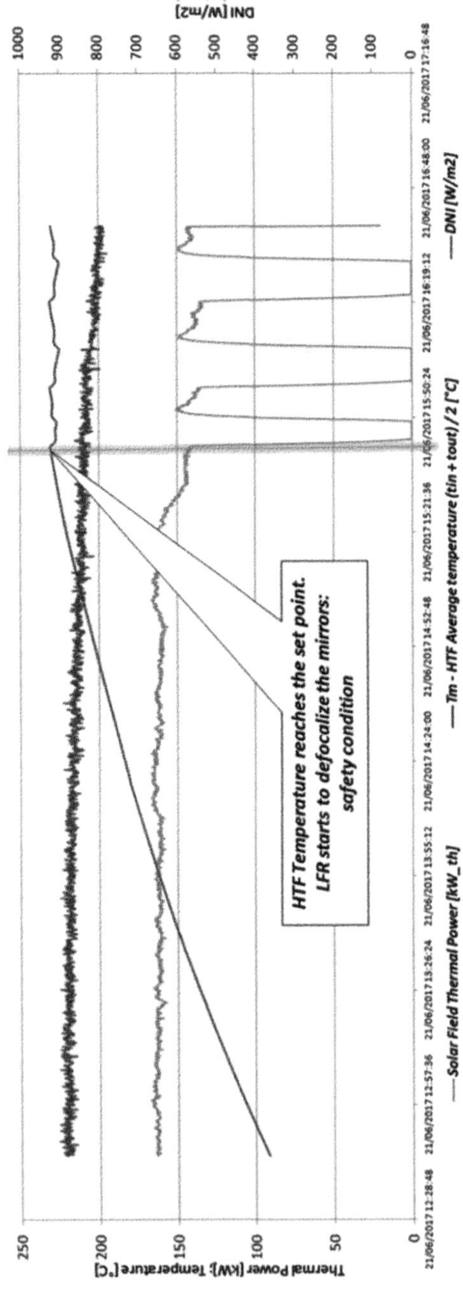

Figure 8.19. *Thermal energy in the solar field with a safety overheating temperature set at 230°C*

The plant in Palermo is equipped with a Li/Br double effect absorption chiller with a heating capacity of 23 kW, at a thermal COP of about 5.1. The chiller, driven by oil, is heated at 195°C and the outlet temperature is 165°C. It is integrated with a cooling tower, which keeps the cooled water at 7°C to 12°C.

8.2.2.6. *The ORC system*

The ORC generator installed in Palermo has a net electric power output of 10 kW. The basic concept of its operation is shown in Figure 8.16 and the operational characteristics of the ORC are depicted in Figure 8.17.

The heat transfer fluid passes its thermal energy to the high temperature working fluid vapor in the evaporator. The high temperature fluid vapor enters the turbine and produces mechanical work for electricity generation. The turbine exhaust steam is condensed in the condenser and pumped again to the evaporator for recirculation (Carlqvist 2009). The main characteristics of the ORC cycle working fluid are thermal stability, toxicity, flammability, and cost. As an example, toluene has good technical characteristics regarding stability, temperature range, and high energy density, but condenses at sub-atmospheric pressures. The n-pentane, on the other hand, operates at super-atmospheric pressures over a wide range of condensing conditions.

Several fluids could be exploited in the ORC generation process, and the suitability of each alternative option depends on the following aspects:

– the location of the power plant. If it is located near a populated or environmentally sensitive area, more flammable and toxic fluids might be excluded and less hazardous mediums are preferred;

– the environmental aspects, that is, the fluid global warming potential;

– the cost of the fluid. The best performing fluid might not be the one with the lowest procurement cost.

The ORC generator installed in Palermo uses a heavy fluid R245fa that allows the inlet temperature to rise up to 210°C and the outlet oil temperature to 165°C. An external cooling tower of 90 kW_{th} is connected in order to discard the waste heat.

8.2.2.7. *Performance analysis of the energy storage of the CSP*

8.2.2.7.1. Quasi dynamic state calculation of LFR energy production

A detailed hourly steady-state model is developed using TRNSYS, taking into consideration the climatic data and the geographic location of the installation site, as well as the optical characteristics of the solar thermal plant. The thermal model is based on the international standard ISO 9806.

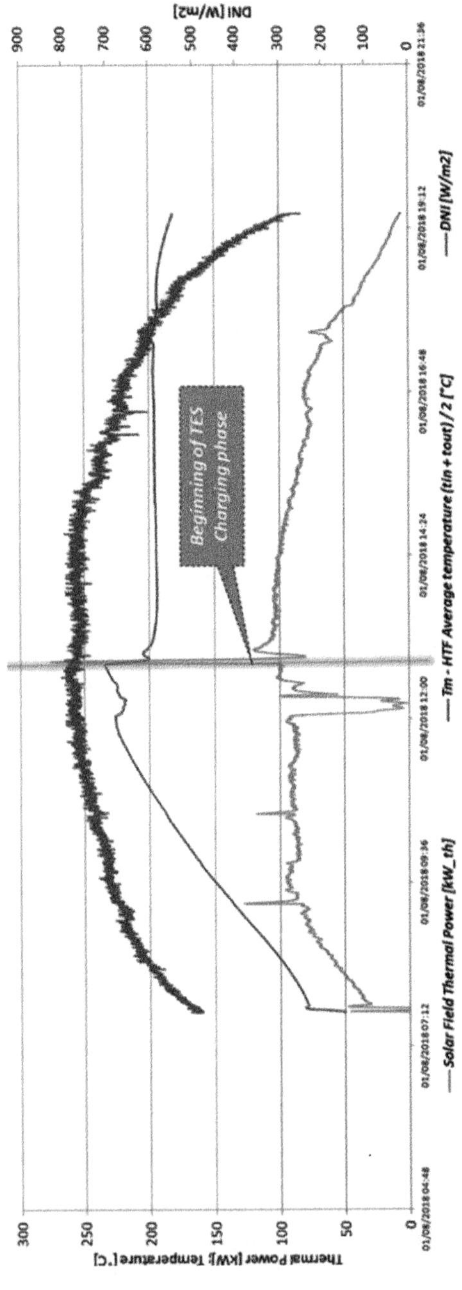

Figure 8.20. *Thermal energy produced in the solar field and its modification during the charging phase of the MS TES*

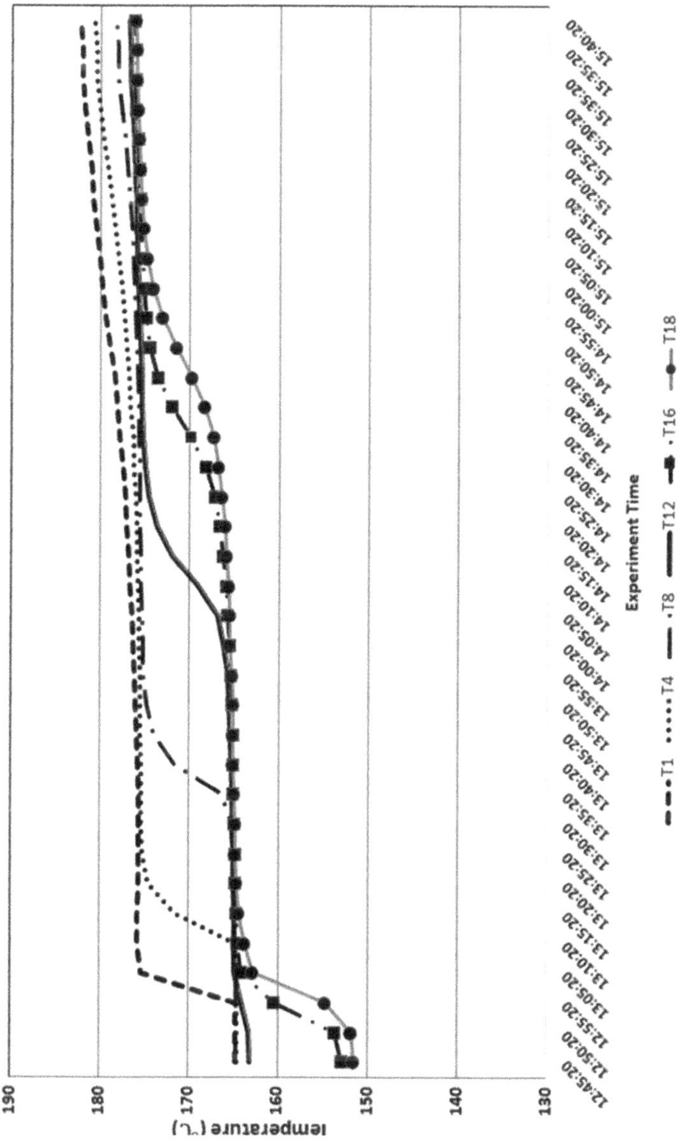

Figure 8.21. *Thermocline effect measured by specific thermal sensors (PT100) distributed in the vertical axis of the thermal storage*

Taking into account the solar radiation and ambient temperatures from the nearest weather data site in the University of Palermo, and a temperature difference of Tin = 270°C and Tout = 250°C, with an optical efficiency of 0.64, the monthly thermal energy output is calculated as shown in Figure 8.18. The results show an annual LFR energy production of 295 MWh (Table 8.5).

8.2.2.7.2. Monitoring of the solar field

The solar field is monitored by a SCADA system connected to all of the sensors distributed in the circuit, in order to control and manage the main working parameters. Figure 8.19 shows the registered temperatures and thermal power of the LFCs installed in Palermo during the test on June 21. The average temperature of the HTF (i.e. Paratherm NF) between the inlet and outlet of the absorber tubes of the three rows is represented in red. The direct radiation on the normal plane is in blue and is measured in W/m_2. The total power gained by the solar field during the test is represented in orange.

During the test, the solar field produced about 550 kWh_{th} (considering the thermal losses of distribution) in almost four hours. The thermal power reached a peak greater than 170 kW_{th}, with a thermal efficiency of 45%, corresponding to an optical peak efficiency of about 50%. This value is acceptable considering the reduced reflectance of mirrors and transmittance of the borosilicate glass of the absorber tubes due to the dust. Safety systems, implemented in the solar field, verify the set point temperatures achievements.

When set points are reached the mirrors are defocused, as depicted in Figure 8.20.

Figure 8.21 shows the performance of LFR while TES is in the charging phase. The average HTF temperature is increased to reach the predefined set point. Then, when the oil is pumped from the buffer to the charging heat exchanger, thermal power is stored as represented in Figure 8.20. Thereafter, the HTF temperature remains almost constant due to the continuous energy transfer from the solar field to the thermal storage. The solar field has produced, at the described operative conditions, 880 kW_{th}, of which 550kW_{th} have been generated during the thermal charging phase of the molten salts.

The thermal energy transferred to the storage system produces a thermocline effect, even at low-temperature thermal profiles. Figure 8.21 shows the thermocline stratification produced by oil pumped at 200°C into the TES, with molten salts at about 160°C. During the charging experiment, the temperature profiles, measured at different heights by the sensors from 1 (top) to 20 (bottom), are progressively increased, while maintaining a sufficient stratification.

The thermal energy stored is used for driving both the absorption chiller and the ORC turbine.

In this specific plant configuration above a threshold of 210°C in the heat discharging zone, the thermal energy can be transferred to the ORC. In the experiment, it has been recorded that with an oil flow of 3.2 m^3/hour and an inlet temperature of 176.2°C and an outlet temperature of 159.7°C, the ORC has a power of 8.8 kW_e. Under these conditions the evaporative tower was fed with 7 m^3/hour of water at 34.6°C, cooling the ORC discharge at 30.1°C (Figure 8.22).

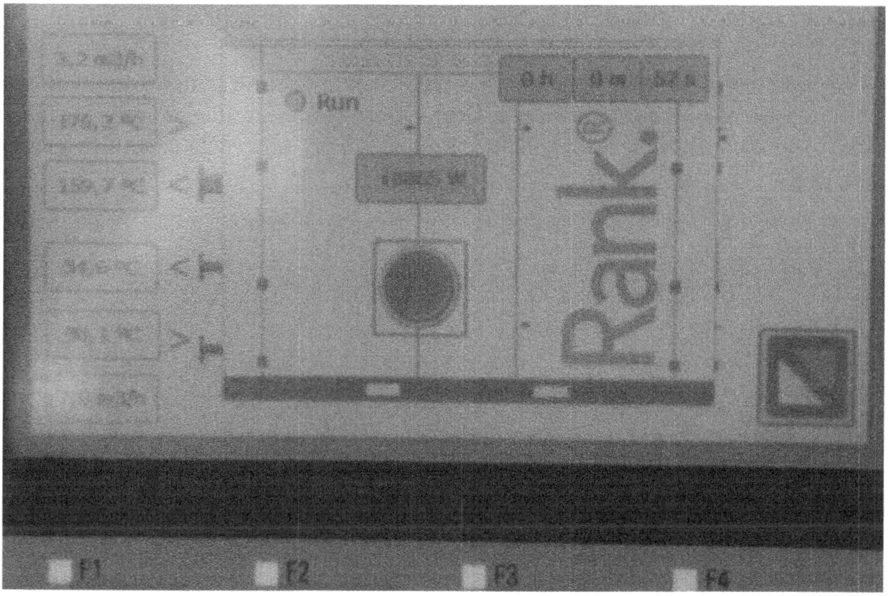

Figure 8.22. *Electric power produced by the rank organic Rankine cycle (ORC) by thermal energy stored in the thermal energy storage system (TES)*

8.3. Conclusion and future prospects

Smart grids are expanded in districts, communities, and cities in order to meet the reduction of energy requirements and achieve sustainable and resilient energy supply. In this smart grids' era energy storage systems play a major and critical role in both energy supply and demand. Peak power shaving, load shifting, demand response, and dynamic pricing can be effectively supported with the appropriate and careful selection of energy storage. Electrical and electrochemical energy storage technologies are the first choices when considering smart grids. EVs technology is steadily enhanced, usually becoming one of the dominant solutions for smart communities and smart cities. On the other hand, batteries are expected to be significantly upgraded, while advanced control systems and devices can ensure that the battery's life reaches its maximum.

Although emphasis is put on electricity supply and demand and its coverage by renewable energy systems, polygeneration and thermal energy storage are equally important for sustainable and resilient energy systems. Significant research effort is put toward advanced thermal energy storage systems with increased capacity and efficiency. Ground source heat pumps, intelligent materials with improved thermal properties are continuously developed and integrated at the community and city level.

Smart, resilient cities of the future should be able to adapt to new conditions and thrive, rather than merely survive. In the roadmap toward resilience, polygeneration and "polystorage" will ensure the security of supply and high quality of energy services.

Conclusion and Recommendations

The development of smart grids worldwide offers unique opportunities with respect to the decarbonization of the building sector. Smart grids promote sophisticated and distributed energy management while utilizing a high level of intelligence in balancing energy production and consumption. This includes significant capabilities for exploiting the necessary very high penetration levels of renewable energy production, in terms of demand side management and the variety of the available energy storage applications.

The EU energy efficiency policies for buildings have been a reference point worldwide over the last half century, leading to a holistic approach to buildings. The EU is establishing a new, diverse spectrum of consecutive strategies concerning energy efficiency in the building sector, such as the renovation wave, and is pursuing ambitious climate-related goals, such as those identified in the EU Green Deal (Economidou *et al.* 2020). These efforts need to be strengthened and closely monitored in order for several challenges to be overcome and for the EU to stimulate change and bridge the gap in order to meet the targets of the Paris Agreement.

The Energy Efficiency First principle and the new European Bauhaus paradigm are drawing great attention with respect to the design and construction of buildings. In addition, a holistic perspective requires encompassing a lifecycle approach and a cyclical economic model in order to account not only for the operational energy, provided it is minimized and covered by renewables, but also for the embodied energy and associated emissions in building materials, components and processes.

In this context, integrated design provides the methodologies and tools for setting and fulfilling complementary targets at the stage of the design and construction of buildings. This entails making decisions based on the combined

Conclusion written by Nikos KAMPELIS.

knowledge brought by the various specialties and roles in a co-creative collaborative fashion towards an efficient use of capital investment and natural resources, in order to establish optimal smart and zero energy building designs.

On the other hand, abiding by zero energy targets in the existing building stock requires retrofitting, either in relation to the implementation of energy efficiency measures (i.e. improvement of the building envelope or installed systems), or in relation to the integration of renewable energy systems and storage equipment. In this regard, the systematic collection of data, data analysis and energy modeling, prior to the interventions, is of great importance as it allows one to identify energy, emissions and cost savings through the design of customized solutions that will be suitable in terms of case-specific objectives and constraints. Also, post-monitoring allows one to validate the expected savings versus the actual ones, otherwise refered to as the performance gap. Overall, monitoring provides the necessary basis for continuous improvements toward the accomplishment of zero net energy consumption or zero net emissions whereby very low energy requirements are met exclusively by the generation of renewable energy.

On the other hand, consumer awareness and a rational use of energy through behavioral change is required together with smart energy management at the building and district levels. This calls for interoperability, secure bi-directional use of dynamic information as well as more recent tools and services that will provide flexibility and reliability along with financial motives for a just transition. Furthermore, meeting clean energy-associated goals in buildings cannot be done in isolation of indoor environmental quality and satisfaction related to users' basic needs. In this sense, thermal comfort and indoor environmental quality are essential for the well-being of a building's occupants and need to be assessed and maintained at certain levels or even be improved. Indoor conditions are very much related to human health, productivity and creativity, therefore they concern a wide spectrum of scientific disciplines including engineering, medical and social sciences. For instance, thermal comfort sensor data can be complemented by subjective analysis to collectively assess the users' perception about how they feel, which may depend on many factors including ones that are physiological, social, ergonomical, practical etc.

Moreover, occupancy is another crucial parameter which needs to be considered alongside dealing with reducing energy consumption, especially in large buildings. Occupancy data may be linked on the one hand to the scheduling, reorganization and reallocation of activities and services provided in a building or in a building complex, such as in universities, hospitals, hotels, airports etc., and can on the other hand be used to embed intelligence in the energy management of building facilities.

In addition, the knowledge, perceptions and behaviors of occupants and their engagement in the effort for a sustainably built environment is a key factor that is gradually receiving higher attention. It is a fact that the knowledge of building users about building systems controls has an important influence on how these are being used. In many cases, limited understanding about systems' operation and "the way things are done" translates to wasting energy and low levels of comfort. Therefore, engaging building users and occupants in the effort to use energy in a sustainable way, is in itself important for accumulating environmental benefits. This can be achieved by raising awareness about the impact of energy use for the environment and for peoples' health, by providing proper training and advice and by developing the right tools that will allow better interaction and the shift to more responsible behaviour.

Aligned with the aforementioned remarks, the adoption of smart and renewable energy communities is proven to be a technically and financially viable option, while at the same time being highly beneficial in empowering social cohesion, respecting cultural values, caring for the environment and human health, both locally and globally. This is supported by numerous literature studies and elaborated also in the recast of the Renewable Energy Directive 2018/2001/EU and the Directive on common rules for the internal electricity market (EU) 2019/44; two high level policy instruments which call EU MS to strengthen the role of the citizen and renewable energy communities in producing, consuming, storing and selling renewable energy, or by providing flexibility services through demand response and storage. In this way, the EU is opening new frontiers for citizens, local authorities and various collaborative schemes to organize in novel ways and become active in the energy transition, as well as participate in electricity markets. Therefore, an important window of opportunity for local societies, political leaders and other key stakeholders is to embrace the momentum and support the creation of ecosystems capable of addressing technological, social and environmental challenges. Next to the implementation of well-established solutions, ecosystems should be given the space and resources to flourish and bring life to social innovations based on the shared values of inclusion, democratic participation and solidarity for vulnerable groups of the society.

References

AIDA Project (2013). AIDA – Affirmative Integrated Energy Design Action [Online]. Available at: https://www.aidaproject.eu/?lng=gr.

Aktas, A., Erhan, K., Özdemir, S., Özdemir, E. (2018). Dynamic energy management for photovoltaic power system including hybrid energy storage in smart grid applications. *Energy*, 162, 72–82.

Alotto, P., Guarnieri, M., Moro, F. (2014). Redox flow batteries for the storage of renewable energy: A review. *Renewable and Sustainable Energy Reviews*, 29, 325–335.

Alva, G., Lin, Y., Fang, G. (2018). An overview of thermal energy storage systems. *Energy*, 144, 341–378.

Amin, S.M. and Wollenberg, B.F. (2005). Toward a smart grid: Power delivery for the 21st century. *IEEE Power and Energy Magazine*, 3(5), 34–41.

Angelini, G., Lucchini, A., Manzolini, G. (2014). Comparison of thermocline molten salt storage performances to commercial two-tank configuration. *Energy Procedia*, 49, 694–704.

Anilkumar, T.T., Simon, S.P., Padhy, N.P. (2017). Residential electricity cost minimization model through open well-pico turbine pumped storage system. *Appl. Energy*, 195, 23–35.

ANSI/SHRAE (2012). ANSI/SHRAE standard 135-2010: A data communication protocol for building automation and control network [Online]. Available at: https://webstore.ansi.org/preview-pages/ASHRAE/preview_ANSI+ASHRAE+Standard+135-2010.pdf.

ASHRAE (2010a). ASHRAE STANDARD 55-2010: Thermal environmental conditions for human occupancy [Online]. Available at: http://arco-hvac.ir/wp-content/uploads/2015/11/ASHRAE-55-2010.pdf.

ASHRAE (2010b). Performance measurement protocols for commercial buildings. Report, ASHRAE/CIBSE/USGBC, Atlanta.

Atmaja, T.D. and Amin (2015). Energy storage system using battery and ultracapacitor on mobile charging station for electric vehicle. *Energy Procedia*, 68, 429–437.

Ayar, M., Obuz, S., Trevizan, R.D., Bretas, A.S., Latchman, H.A. (2017). A distributed control approach for enhancing smart grid transient stability and resilience. *IEEE Trans. Smart Grid*, 8(6), 3035–3044.

Bae, M., Kim, H., Kim, E., Chung, A.Y., Kim, H., Roh, J.H. (2014). Toward electricity retail competition: Survey and case study on technical infrastructure for advanced electricity market system. *Appl. Energy*, 133, 252–273.

Bahmani-Firouzi, B. and Azizipanah-Abarghooee, R. (2014). Optimal sizing of battery energy storage for micro-grid operation management using a new improved bat algorithm. *Int. J. Electr. Power Energy Syst.*, 56, 42–54.

Bartels, G. (2012). Global Smart Grid Federation Report [Online]. Available at: https://www.smartgrid.gov/files/documents/Global_Smart_Grid_Federation_Report.pdf.

Bartusch, C. and Alvehag, K. (2014). Further exploring the potential of residential demand response programs in electricity distribution. *Appl. Energy*, 125, 39–59.

Bhaskar Naik, M., Kumar, P., Majhi, S. (2019). Smart public transportation network expansion and its interaction with the grid. *Int. J. Electr. Power Energy Syst.*, 105, 365–380.

Bourgeois, D., Reinhart, C., Macdonald, I. (2006). Adding advanced behavioural models in whole building energy simulation: A study on the total energy impact of manual and automated lighting control. *Energy Build.*, 38(7), 814–823.

Bradley, P., Leach, M., Torriti, J. (2013). A review of the costs and benefits of demand response for electricity in the UK. *Energy Policy*, 52, 312–327.

Burch, J., Subbarao, K., Lekov, A., Warren, M., Norford, L. (1990). Short-term energy monitoring in a large commercial building in ASHRAE transactions. *ASHRAE*, 90, 1459–1477.

Carlisle, N., Otto, A., Geet, V., Pless, S. (2009). Definition of a "Zero Net Energy" Community. Technical report, National Renewable Energy Laboratory [Online]. Available at: https://www.nrel.gov/docs/fy10osti/46065.pdf.

Carlqvist, A. (2009). Overview of concentrating solar power for electricity production, with emphasis on steam turbine aspects. Report, KTH Royal Institute of Technology, Stockholm.

Carpinelli, G., Celli, G., Mocci, S., Mottola, F., Pilo, F., Proto, D. (2013). Optimal integration of distributed energy storage devices in smart grids. *IEEE Trans. Smart Grid*, 4(2), 985–995.

Cazzaniga, R., Cicu, M., Rosa-Clot, M., Rosa-Clot, P., Tina, G.M., Ventura, C. (2017). Compressed air energy storage integrated with floating photovoltaic plant. *J. Energy Storage*, 13, 48–57.

CEN (2005). EN ISO 7730 Ergonomics of the thermal environment – Analytical determination and interpretation of thermal comfort using calculation of the PMV and PPD indices and local thermal comfort criteria (ISO 7730:2005) [Online]. Available at: https://www.iso.org/standard/39155.html.

Chalkiadakis, G., Robu, V., Kota, R., Rogers, A., Jennings, N.R. (2011). Cooperatives of distributed energy resources for efficient virtual power plants. *Proceedings of the 10th International Conference on Autonomous Agents and Multiagent Systems*, 787–794.

Chaudhary, P. and Rizwan, M. (2018). Energy management supporting high penetration of solar photovoltaic generation for smart grid using solar forecasts and pumped hydro storage system. *Renew. Energy*, 118, 928–946.

Chen, H., Cong, T.N., Yang, W., Tan, C., Li, Y., Ding, Y. (2009). Progress in electrical energy storage system: A critical review. *Progress in Natural Science*, 19(3), 291–312.

Chen, L., Huang, Z.C., Wu, P., Liu, C. (2013). Transmission line condition monitoring based on internet of things in smart grid. *Applied Mechanics and Materials*, 336–338, 2488–2493.

Cheng, J. and Choobineh, F. (2016). Microgrid expansion through a compressed air assisted wind energy system. *2016 IEEE Power and Energy Society Innovative Smart Grid Technologies Conference (ISGT)*, September. DOI: 10.1109/ISGT.2016.7781257.

Colak, I., Sagiroglu, S., Fulli, G., Yesilbudak, M., Covrig, C.F. (2016). A survey on the critical issues in smart grid technologies. *Renewable and Sustainable Energy Reviews*, 54, 396–405.

Colmenar-Santos, A., Linares-Mena, A.R., Borge-Diez, D., Quinto-Alemany, C.D. (2017). Impact assessment of electric vehicles on islands grids: A case study for Tenerife (Spain). *Energy*, 120, 385–396.

Colmenar-Santos, A., Molina-Ibáñez, E.L., Rosales-Asensio, E., López-Rey, Á. (2018). Technical approach for the inclusion of superconducting magnetic energy storage in a smart city. *Energy*, 158, 1080–1091.

Crawley, D., Lawrie, L., Winkelmann, F.C., Buhl, W.F., Huang, Y.J., Pedersen, C.O., Strand, R.K., Liesen, R.J., Fisher, D.E., Witte, M.J., Glazer, J. (1999). EnergyPlus: A new-generation building energy simulation program. *Simulation '99*, 1–14.

Das, C.K., Bass, O., Kothapalli, G., Mahmoud, T.S., Habibi, D. (2018). Overview of energy storage systems in distribution networks: Placement, sizing, operation, and power quality. *Renewable and Sustainable Energy Reviews*, 91, 1205–1230.

De Wit, S. and Augenbroe, G. (2002). Analysis of uncertainty in building design evaluations and its implications. *Energy Build.*, 34(9), 951–958.

de Dear, R.J., Leow, K.G., Foo, S.C. (1991). Thermal comfort in the humid tropics: Field experiments in air conditioned and naturally ventilated buildings in Singapore. *Int. J. Biometeorol.*, 34(4), 259–265.

Deng, S., Wang, R.Z., Dai, Y.J. (2014). How to evaluate performance of net zero energy building – A literature research. *Energy*, 71, 1–16.

DoE (2016). DoE Global Energy Storage Database [Online]. Available at: https://sandia.gov/ess-ssl/gesdb/public/.

Dorotić, H., Doračić, B., Dobravec, V., Pukšec, T., Krajačić, G., Duić, N. (2019). Integration of transport and energy sectors in island communities with 100% intermittent renewable energy sources. *Renew. Sustain. Energy Rev.*, 99, 109–124.

Duffy, P., Fitzpatrick, C., Conway, T., Lynch, R.P. (2019). Energy sources and supply grids: The growing need for storage. *Issues Environ. Sci. Technol.*, January(46), 1–41.

Dunn, B., Kamath, H., Tarascon, J.M. (2011). Electrical energy storage for the grid: A battery of choices. *Science*, 334(6058), 928–935.

E.A. International (2012). EV City casebook: A look at the global electric vehicle movement [Online]. Available at: https://urbanforesight.org/wp-content/uploads/2015/08/EVCity Casebook.pdf.

Earth System Research Lab (2021). NOAA ESRL: Solar position calculator [Online]. Available at: https://www.esrl.noaa.gov/gmd/grad/ solcalc/azel.html.

Economidou, M., Todeschi, V., Bertoldi, P., D'Agostino, D., Zangheri, P., Castellazzi, L. (2020). Review of 50 years of EU energy efficiency policies for buildings. *Energy and Buildings*, 225, 110322.

El-Hawary, M.E. (2014). The smart grid – State-of-the-art and future trends. *Electr. Power Components Syst.*, 42(3–4), 239–250.

Elminir, H.K., Azzam, Y.A., Younes, F.I. (2007). Prediction of hourly and daily diffuse fraction using neural network, as compared to linear regression models. *Energy*, 32(8), 1513–1523.

Erlinghagen, S. and Markard, J. (2012). Smart grids and the transformation of the electricity sector: ICT firms as potential catalysts for sectoral change. *Energy Policy*, 51, 895–906.

Espe, E., Potdar, V., Chang, E. (2018). Prosumer communities and relationships in smart grids: A literature review, evolution and future directions. *Energies*, 11(10), 2528.

European Standardisation Organisation (2007). Indoor environmental input parameters for design and assessment of energy performance of buildings addressing indoor air quality, thermal environment, lighting and acoustics. Standard, CEN EN 15251:2007, European Standardisation Organisation, European Standardisation Organisation, Brussels.

European Union (2019). The European Green Deal EN. Policy, EU, Brussels.

Fan, R.E., Chang, K.W., Hsieh, C.J., Wang, X.R., Lin, C.J. (2008). LIBLINEAR: A library for large linear classification. *J. Mach. Learn. Res.*, 9, 1871–1874.

Fan, C., Xiao, F., Wang, S. (2014). Development of prediction models for next-day building energy consumption and peak power demand using data mining techniques. *Appl. Energy*, 127, 1–10.

Fanos, K. (2014). Energy auditor at the Cyprus Institute Novel Technologies Laboratory. Nicosia.

Farraj, A., Hammad, E., Kundur, D. (2018). A storage-based multiagent regulation framework for smart grid resilience. *IEEE Trans. Ind. Informatics*, 14(9), 3859–3869.

Fedak, W., Anweiler, S., Ulbrich, R., Jarosz, B. (2017). The concept of autonomous power supply system fed with renewable energy sources. *J. Sustain. Dev. Energy, Water Environ. Syst.*, 5(4), 579–589.

Finck, C., Li, R., Zeiler, W. (2015). An optimization strategy for scheduling various thermal energy storage technologies in office buildings connected to smart grid. *Energy Procedia*, 78, 806–811.

Fischer, D. and Madani, H. (2017). On heat pumps in smart grids: A review. *Renewable and Sustainable Energy Reviews*, 70, 342–357.

Flores, J.F.C., Espagnet, A.R., Chiu, J.N.W., Martin, V., Lacarrière, B. (2017). Techno-economic assessment of active latent heat thermal energy storage systems with low-temperature district heating. *Int. J. Sustain. Energy Plan. Manag.*, 13, 5–18.

Fokaides, P.A., Maxoulis, C.N., Panayiotou, G.P., Neophytou, M.K.-A., Kalogirou, S.A. (2011). Comparison between measured and calculated energy performance for dwellings in a summer dominant environment. *Energy Build.*, 43(11), 3099–3105.

Gaiser, K. and Stroeve, P. (2014). The impact of scheduling appliances and rate structure on bill savings for net-zero energy communities: Application to West Village. *Appl. Energy*, 113, 1586–1595.

Garg, V. and Bansal, N.K. (2000). Smart occupancy sensors to reduce energy consumption. *Energy Build.*, 32(1), 81–87.

Ghiani, E., Mocci, S., Celli, G., Pilo, F. (2014). Increasing the flexible use of hydro pumping storage for maximizing the exploitation of RES in Sardinia. *3rd Renewable Power Generation Conference, (RPG 2014)*, 1–6 [Online]. Available at: doi: 10.1049/cp.2014.0853.

Gibb, D., Seitz, A., Johnson, M., Romani, J., Gasia, J., Cabeza, L.F., Gurtner, R. (2018). Applications of thermal energy storage in the energy transition – Benchmarks and developments. Public report, IEA ECES Annex 30.

Giessler, P., Steinegger, R., Abeck, S., Gebhart, M. (2016). Checklist for the API design of web services based on REST. *Int. J. Adv. Internet Technol.*, 4, 41–51.

Giordano V., Meletiou A., Covrig C.F., Mengolini A., Ardelean M., Fulli G. (2013). Smart Grid projects in Europe: Lessons learned and current developments 2012 update. JRC, European Commission [Online]. Available at: https://ses.jrc.ec.europa.eu/publications/reports/smart-grid-projects-europe-lessons-learned-and-current-developments-2012-update.

Gobakis, K., Kolokotsa, D., Synnefa, A., Saliari, M., Giannopoulou, K., Santamouris, M. (2011). Development of a model for urban heat island prediction using neural network techniques. *Sustain. Cities Soc.*, 1(2), 104–115.

Gobakis, K., Mavrigiannaki, A., Kalaitzakis, K., Kolokotsa, D.D. (2017). Design and development of a Web based GIS platform for zero energy settlements monitoring. *Energy Procedia*, 134, 48–60.

Godina, R., Rodrigues, E.M.G., Matias, J.C.O., Catalão, J.P.S. (2016). Smart electric vehicle charging scheduler for overloading prevention of an industry client power distribution transformer. *Appl. Energy*, 178, 29–42.

González-Roubaud, E., Pérez-Osorio, D., Prieto, C. (2017). Review of commercial thermal energy storage in concentrated solar power plants: Steam vs. molten salts. *Renewable and Sustainable Energy Reviews*, 80, 133–148.

Graditi, G., Ippolito, M.G., Telaretti, E., Zizzo, G. (2016). Technical and economical assessment of distributed electrochemical storages for load shifting applications: An Italian case study. *Renewable and Sustainable Energy Reviews*, 57, 515–523.

Guarino, F., Athienitis, A., Cellura, M., Bastien, D. (2017). PCM thermal storage design in buildings: Experimental studies and applications to solaria in cold climates. *Appl. Energy*, 185, 95–106.

Güngör, V.C., Sahim, D., Kocak, T., Ergut, S., Buccella, C., Cecati, C., Hancke, G.P. (2011). Smart grid technologies: Communication technologies and standards. *IEEE Trans. Ind. Informatics*, 7(4), 529–539.

Haberl, J.S. and Komor, P.S. (1990). Improving energy audits: How annual and monthly consumption data can help. United States [Online]. Available at: https://www.osti.gov/biblio/6555147-improving-energy-audits-how-annual-monthly-consumption-data-can-help.

Haillot, D., Poupin, L., Urrutti, P., Lenôtre, C., Wohrer, M. (2011). Low temperature solar power plant connected to a smart grid. *ISES Solar World Congress*, 5, 3920–3925. [Online]. Available at: https://www.researchgate.net/publication/314884380_Low_Temperature_Solar_Power_Plant_Connected_to_a_Smart_Grid.

Hartmann, N., Vöhringer, O., Kruck, C., Eltrop, L. (2012). Simulation and analysis of different adiabatic Compressed Air Energy Storage plant configurations. *Appl. Energy*, 93, 541–548.

Hearn, C.S., Lewis, M.C., Pratap, D., Hebner, R., Uriarte, F., Chen, D., Longoria, R. (2013). Utilization of optimal control law to size grid-level flywheel energy storage. *IEEE Trans. Sustain. Energy*, 4(3), 611–618.

Hestnes, A.G. and Kofoed, N.U. (2002). Effective retrofitting scenarios for energy efficiency and comfort: Results of the design and evaluation activities within the OFFICE project. *Build. Environ.*, 37(6), 569–574.

Hitchin, R. and Knight, I. (2016). Daily energy consumption signatures and control charts for air-conditioned buildings. *Energy Build.*, 112, 101–109.

Hodo-Abalo, S., Banna, M., Zeghmati, B. (2012). Performance analysis of a planted roof as a passive cooling technique in hot-humid tropics. *Renew. Energy*, 39(1), 140–148.

Hong, S.H., Kim, S.H., Kim, G.M., Kim, H.L. (2014). Experimental evaluation of BZ-GW (BACnet-ZigBee smart grid gateway) for demand response in buildings. *Energy*, 65, 62–70.

Hutson, C., Venayagamoorthy, G.K., Corzine, K.A. (2008). Intelligent scheduling of hybrid and electric vehicle storage capacity in a parking lot for profit maximization in grid power transactions. *Proceedings of the IEEE Energy 2030 Conference, ENERGY 2008*, Institute of Electrical and Electronics Engineers, Atlanta.

ICE-WISH Project (n/a). The Common ICT-PSP Methodology for Impact Assessment. Grant Agreement number 270898.

IEC (2011). IEC White Paper EES:2011. Electrical Energy Storage. White Paper, IEC.

ISO (2005). ISO 7730: Ergonomics of the thermal environment. Analytical determination and interpretation of thermal comfort using calculation of the PMV and PPD indices and local thermal comfort criteria. *Management*, 3(7–10), 605–615.

Jain, R.K., Smith, K.M., Culligan, P.J., Taylor, J.E. (2014). Forecasting energy consumption of multi-family residential buildings using support vector regression: Investigating the impact of temporal and spatial monitoring granularity on performance accuracy. *Appl. Energy*, 123, 168–178.

Janotte, N., Meiser, S., Krüger, D., Lüpfert, E., Pitz-Paal, R., Fischer, S., Müller-Steinhagen, H. (2009). Quasi-dynamic analysis of thermal performance of parabolic trough collectors. *SolarPACES 2009 Conference Proceedings*, 15–18 September. Berlin [Online]. Available at: https://elib.dlr.de/61659/.

Javed, F., Arshad, N., Wallin, F., Vassileva, I., Dahlquist, E. (2012). Forecasting for demand response in smart grids: An analysis on use of anthropologic and structural data and short term multiple loads forecasting. *Appl. Energy*, 96, 150–160.

Jetcheva, J.G., Majidpour, M., Chen, W.P. (2014). Neural network model ensembles for building-level electricity load forecasts. *Energy Build.*, 84, 214–223.

Joaquim, P.C.E. and Rosa, J.L.G. (2005). Artificial neural networks for temporal processing applied to prediction of electric energy generation in small hydroelectric power stations. *Proceedings of the International Joint Conference on Neural Networks*, 4, 2625–2630.

Kalogirou, S.A. (2004). Solar thermal collectors and applications. *Prog. Energy Combust. Sci.*, 30(3), 231–295.

Kapsalaki, M. and Leal, V. (2011). Recent progress on net zero energy buildings. *Adv. Build. Energy Res.*, 5(1), 129–162.

Karatasou, S., Santamouris, M., Geros, V. (2006). Modeling and predicting building's energy use with artificial neural networks: Methods and results. *Energy Build.*, 38(8), 949–958.

Karlessi, T., Amann, S., Nordby, A.S., Leutgöb, K. (2014). Adapting the principles of Integrated Design to achieve high performance goals: Nearly Zero Energy Building in the European market. *8th International Conference Improving Energy Efficiency in Commercial Buildings (IEECB'14)*.

Karlessi, T., Kampelis, N., Kolokotsa, D., Santamouris, M., Standardi, L., Isidori, D., Cristalli, C. (2017). The concept of smart and NZEB buildings and the integrated design approach. *Procedia Engineering*, 180, 1316–1325.

Katafygiotou, M.C. and Serghides, D.K. (2014). Thermal comfort of a typical secondary school building in Cyprus. *Sustainable Cities and Society*, 13, 303–312 [Online]. Available at: https://www.sciencedirect.com/science/article/pii/S2210670714000250.

Kayo, G., Hasan, A., Siren, K. (2014). Energy sharing and matching in different combinations of buildings, CHP capacities and operation strategy. *Energy Build.*, 82, 685–695.

Keerin, P., Kurutach, W., Boongoen, T. (2012). Cluster-based kNN missing value imputation for DNA microarray data. In *Conference Proceedings – IEEE International Conference on Systems, Man and Cybernetics*, 445–450.

Keles, C., Karabiber, A., Akcin, M., Kaygusuz, A., Alagoz, B.B., Gul, O. (2015). A smart building power management concept: Smart socket applications with DC distribution. *Int. J. Electr. Power Energy Syst.*, 64, 679–688.

Kephart, J.O. and Chess, D.M. (2003). The vision of autonomic computing. *Computer (Long. Beach. Calif).*, 36(1), 41–50.

Khan, A.R., Mahmood, A., Safdar, A., Khan, Z.A., Khan, N.A. (2016). Load forecasting, dynamic pricing and DSM in smart grid: A review. *Renewable and Sustainable Energy Reviews*, 54, 1311–1322.

Kohlhepp, P. and Hagenmeyer, V. (2017). Technical potential of buildings in Germany as flexible power-to-heat storage for smart-grid operation. *Energy Technol.*, 5(7), 1084–1104.

Kolb, G.J. (2011). Evaluation of annual performance of 2-tank and thermocline thermal storage systems for trough plants. *J. Sol. Energy Eng. Trans. ASME*, 133(3), 1–5.

Kolokotsa, D., Stavrakakis, G.S., Kalaitzakis, K., Agoris, D. (2002). Genetic algorithms optimized fuzzy controller for the indoor environmental management in buildings implemented using PLC and local operating networks. *Eng. Appl. Artif. Intell.*, 15(5), 417–428.

Kolokotsa, D., Pouliezos, A., Stavrakakis, G., Lazos, C. (2009). Predictive control techniques for energy and indoor environmental quality management in buildings. *Build. Environ.*, 44(9), 1850–1863.

Kolokotsa, D., Rovas, D., Kosmatopoulos, E., Kalaitzakis, K. (2011). A roadmap towards intelligent net zero- and positive-energy buildings. *Solar Energy*, 85(12), 3067–3084.

Kolokotsa, D., Provata, E., Papantoniou, S., Pietrini, M., Giovannelli, A., Romiti, G. (2015). Development of optimization algorithms for the Leaf Community microgrid. *Renew. Energy*, 74, 782–795.

Kolokotsa, D., Gobakis, K., Papantoniou, S., Georgatou, C., Kampelis, N., Kalaitzakis, K., Vasilakopoulou, K., Santamouris, M. (2016a). Development of a web based energy management system for University Campuses: The CAMP-IT platform. *Energy Build.*, 123(July), 119–135.

Kolokotsa, D., Kampelis, N., Karlessi, T., Standardi, L., Cristalli, C. (2016b). Smart grid energy management staff exchange: D3.2 Guidelines for smart buildings and smart grids. Report on the operational phase, SmartGems Energy Network.

Koohi-Kamali, S., Tyagi, V.V., Rahim, N.A., Panwar, N.L., Mokhlis, H. (2013). Emergence of energy storage technologies as the solution for reliable operation of smart power systems: A review. *Renewable and Sustainable Energy Reviews*, 25, 135–165.

Kwong, Q.J., Adam, N.M., Sahari, B.B. (2014). Thermal comfort assessment and potential for energy efficiency enhancement in modern tropical buildings: A review. *Energy and Buildings*, 68(PARTA), 547–557.

Kyriakarakos, G., Dounis, A.I., Arvanitis, K.G., Papadakis, G. (2012). A fuzzy logic energy management system for polygeneration microgrids. *Renew. Energy*, 41, 315–327.

Lam, K.P., Huang, Y.C., Zhai, C. (2004). Energy modeling tools assessment for early design phase. Center for Building Performance and Diagnostics/School of Architecture Carnegie Mellon University.

Langevin, J., Wen, J., Gurian, P.L. (2016). Quantifying the human-building interaction: Considering the active, adaptive occupant in building performance simulation. *Energy Build.*, 117, 372–386.

Latha, R., Palanivel, S., Kanakaraj, J. (2012). Frequency control of microgrid based on compressed air energy storage system. *Distrib. Gener. Altern. Energy J.*, 27(4), 8–19.

Leutgöb, K. (2012). Annex 1 Description of the Action. Intelligent Energy Europe – MaTrID project [Online]. Available at: http://www.integrateddesign.eu.

Leutgöb, K. (2013). MaTrID: Market transformation towards nearly zero energy buildings through widespread use of integrated energy design [Online]. Available at: http://www.integrateddesign.eu.

Li, G. (2016). Sensible heat thermal storage energy and exergy performance evaluations. *Renewable and Sustainable Energy Reviews*, 53, 897–923.

Li, X.H. and Hong, S.H. (2014a). User-expected price-based demand response algorithm for a home-to-grid system. *Energy*, 64, 437–449.

Li, Y.C. and Hong, S.H. (2014b). BACnet-EnOcean Smart Grid Gateway and its application to demand response in buildings. *Energy Build.*, 78, 183–191.

Li, D.H.W., Yang, L., Lam, J.C. (2013). Zero energy buildings and sustainable development implications: A review. *Energy*, (C), 541–10.

Liang, H.Z. and Gooi, H.B. (2010). Unit commitment in microgrids by improved genetic algorithm. *2010 9th International Power and Energy Conference, IPEC 2010*, 842–847.

Lucas, A. and Chondrogiannis, S. (2016). Smart grid energy storage controller for frequency regulation and peak shaving, using a vanadium redox flow battery. *Int. J. Electr. Power Energy Syst.*, 80, 26–36.

Lund, H., Andersen, A.N., Østergaard, P.A., Mathiesen, B.V., Connolly, D. (2012). From electricity smart grids to smart energy systems – A market operation based approach and understanding. *Energy*, 42(1), 96–102.

Luo, Y., Zhu, T., Wan, S., Zhang, S., Li, K. (2016). Optimal charging scheduling for large-scale EV (electric vehicle) deployment based on the interaction of the smart-grid and intelligent-transport systems. *Energy*, 97, 359–368.

Mahdavi, A., Mohammadi, A., Kabir, E., Lambeva, L. (2008). Occupants' operation of lighting and shading systems in office buildings. *J. Build. Perform. Simul.*, 1(1), 57–65.

Maleki, H., Khederzadeh, M., Asgharian, V. (2014). Local control of active and reactive power in inverter based micro grids. *IEEE PES Innovative Smart Grid Technologies*, Europe, 1–6 [Online]. Available at: doi: 10.1109/ISGTEurope.2014.7028844.

Martha, C., Katafygiotou, D., Serghides, K. (2014). Thermal comfort of a typical secondary school building in Cyprus, Sustainable Cities and Society, 13, 303–312 [Online]. Available at: https://doi.org/10.1016/j.scs.2014.03.004 and https://www.sciencedirect.com/science/article/pii/S2210670714000250.

Masy, G., Georges, E., Verhelst, C., Lemort, V., André, P. (2015). Smart grid energy flexible buildings through the use of heat pumps and building thermal mass as energy storage in the Belgian context. *Sci. Technol. Built Environ.*, 21(6), 800–811.

May, G.J., Davidson, A., Monahov, B. (2018). Lead batteries for utility energy storage: A review. *Journal of Energy Storage*, 15, 145–157.

van Meerwijk, A.J.H., Benders, R.M.J., Davila-Martinez, A, Laugs, G.A.H. (2016). Swiss pumped hydro storage potential for Germany's electricity system under high penetration of intermittent renewable energy. *J. Mod. Power Syst. Clean Energy*, 4(4), 542–553.

Mellit, A. and Pavan, A.M. (2010). A 24-h forecast of solar irradiance using artificial neural network: Application for performance prediction of a grid-connected PV plant at Trieste, Italy. *Sol. Energy*, 84(5), 807–821.

Mercier, P., Cherkaoui, R., Oudalov, A. (2009). Optimizing a battery energy storage system for frequency control application in an isolated power system. *IEEE Trans. Power Syst.*, 24(3), 1469–1477.

Meteonorm (n/a). Irradiation data for every place on Earth [Online]. Available at: https://meteonorm.com/en/meteonorm-documents.

Mette, B., Kerskes, H., Drück, H. (2012). Concepts of long-term thermochemical energy storage for solar thermal applications: Selected examples. *Energy Procedia*, 30, 321–330.

Mohamed, F.A. and Koivo, H.N. (2007). Online management of MicroGrid with battery storage using multiobjective optimization. *POWERENG 2007 – International Conference on Power Engineering – Energy and Electrical Drives Proceedings*, 231–236.

Mohamed, F.A. and Koivo, H.N. (2012). Online management genetic algorithms of microgrid for residential application. *Energy Conversion and Management*, 64, 562–568.

Molina, M.G. (2017). Energy storage and power electronics technologies: A strong combination to empower the transformation to the smart grid. *Proc. IEEE*, 105(11), 2191–2219.

Montagnino, F.M. (2017). Solar cooling technologies. Design, application and performance of existing projects. *Sol. Energy*, 154, 144–157.

Montenon, A., Paredes, F., Giaconia, A., Fylaktos, N., Di Bono, S., Papanicolas, C.N., Montagnino, F. (2017). Solar multi-generation in the Mediterranean area, the experience of the STS-MED project. *International Solar Energy Society*, 1–9 [Online]. Available at: http://proceedings.ises.org/paper/eurosun2016/eurosun2016-0092-Montagnino.pdf.

Montero-Izquierdo, A., Hirsch, T., Schenk, H., Bruno, J.C., Coronas, A. (2011). Performance analysis of absorption cooling systems using linear Fresnel solar collectors. *Third International Conference on Applied Energy*, Perugia, Italy, 1199–1210 [Online]. Available at: https://www.researchgate.net/publication/279531278_Performance_Analysis_of_Absorption_Cooling_Systems_Using_Linear_Fresnel_Solar_Collectors.

Morbitzer, C.A. (2003). Towards the integration of simulation into the building design process. PhD thesis, Energy System Research Unit, University of Strathclyde, Glasgow.

Moslehi, K. and Kumar, R. (2010). A reliability perspective of the smart grid. *IEEE Trans. Smart Grid*, 1(1), 57–64.

Mozafar, M.R., Moradi, M.H., Amini, M.H. (2017). A simultaneous approach for optimal allocation of renewable energy sources and electric vehicle charging stations in smart grids based on improved GA-PSO algorithm. *Sustain. Cities Soc.*, 32, 627–637.

Muh, E. and Tabet, F. (2019). Comparative analysis of hybrid renewable energy systems for off-grid applications in Southern Cameroons. *Renew. Energy*, 135, 41–54.

Mukherjee, J.C. and Gupta, A. (2015). A review of charge scheduling of electric vehicles in smart grid. *IEEE Syst. J.*, 9(4), 1541–1553.

Naiyang, G., Dacheng, T., Zhigang, L., Bo, Y. (2012). Online nonnegative matrix factorization with robust stochastic approximation. *IEEE Trans. Neural Networks Learn. Syst.*, 23(7), 1087–1099.

Nefedov, E., Sierla, S., Vyatkin, V. (2018). Internet of energy approach for sustainable use of electric vehicles as energy storage of prosumer buildings. *Energies*, 11(8), 2165.

Nikolaou, T.G., Kolokotsa, D.S., Stavrakakis, G.S., Skias, I.D. (2012). On the application of clustering techniques for office buildings' energy and thermal comfort classification. *IEEE Trans. Smart Grid*, 3(4), 2196–2210.

Nordby, A.S. (2013). ID Process Guide. Intelligent Energy Europe – MaTrID project [Online]. Available at: http://www.integrateddesign.eu.

Oliveira-Lima, J.A., Delgado-Gomes, V., Martins, J.F., Lima, C. (2014). Standard-based service-oriented infrastructure to integrate intelligent buildings in distributed generation and smart grids. *Energy Build.*, 76, 450–458.

Ortega-Fernández, I., Zavattoni, S.A., Rodríguez-Aseguinolaza, J., D'Aguanno, B., Barbato, M.C. (2017). Analysis of an integrated packed bed thermal energy storage system for heat recovery in compressed air energy storage technology. *Appl. Energy*, 205, 280–293.

Pacheco, J.E., Showalter, S.K., Kolb, W.J. (2002). Development of a molten-salt thermocline thermal storage system for parabolic trough plants. *J. Sol. Energy Eng. Trans. ASME*, 124, 124–153.

Panapakidis, I.P., Papadopoulos, T.A., Christoforidis, G.C., Papagiannis, G.K. (2014). Pattern recognition algorithms for electricity load curve analysis of buildings. *Energy Build.*, 73, 137–145.

Pang, C., Aravinthan, V., Wang, X. (2014). Electric vehicles as configurable distributed energy storage in the smart grid. *2014 Clemson University Power Systems Conference*, 1–5.

Pantazaras, A., Lee, S.E., Santamouris, M., Yang, J. (2016). Predicting the CO2 levels in buildings using deterministic and identified models. *Energy Build.*, 127(127), 774–785.

Papantoniou, S., Kolokotsa, D., Kalaitzakis, K. (2015). Building optimization and control algorithms implemented in existing BEMS using a web based energy management and control system. *Energy Build.*, 98, 45–55.

Papantoniou, S., Kolokotsa, D., Kalaitzakis, K., Cesarini, D.N., Cubi, E., Cristalli, C. (2016). Adaptive lighting controllers using smart sensors. *Int. J. Sustain. Energy*, 35(6), 537–553.

Pardo, P., Deydier, A., Anxionnaz-Minvielle, Z., Rougé, S., Cabassud, M., Cognet, P. (2014). A review on high temperature thermochemical heat energy storage. *Renewable and Sustainable Energy Reviews*, 32, 591–610.

Parvania, M., Fotuhi-Firuzabad, M., Shahidehpour, M. (2014). Comparative hourly scheduling of centralized and distributed storage in day-ahead markets. *IEEE Trans. Sustain. Energy*, 5(3), 729–737.

Patel, S.M., Freeman, P.T., Wagner, J.R. (2014). An electrical microgrid-integration of solar panels, compressed air storage, and a micro-cap gas turbine. *ASME 2014 Dynamic Systems and Control Conference, DSCC 2014*, 2.

Pedregosa, F., Varoquaux, G., Gramfort, A., Michel, V., Thirion, B., Grisel, O., Blondel, M., Prettenhofer, P., Weiss, R., Dubourg, V., Vanderplas, J., Passos, A., Cournapeau, D., Brucher, M., Perrot, M., Duchesnay, É. (2011). Scikit-learn: Machine learning in Python [Online]. Available at: https://scikit-learn.org/stable/about.html#citing-scikit-learn.

Pérez-Romero, M., Lozano-Tello, A., Romero-Cadaval, E., Martins, J. (2013). Community and residential energy storage in smart grids. *IFIP Adv. Inf. Commun. Technol.*, 394, 315–322.

Pfleger, N., Bauer, T., Martin, C., Eck, M., Wörner, A. (2015). Thermal energy storage – Overview and specific insight into nitrate salts for sensible and latent heat storage. *Beilstein Journal of Nanotechnology*, 6(1), 1487–1497.

Pisello, A.L., D'Alessandro, A., Fabiani, C., Fiorelli, A.P., Ubertini, F., Cabeza, L.F., Materazzi, A.L., Cotana, F. (2017). Multifunctional analysis of innovative PCM-filled concretes. *Energy Procedia*, 111, 81–90.

Privat, G. (2013). How information and communication technologies will shape SmartGrids. In *Smart Grids*, Hadjsaïd, N. and Sabonnadière, J.-C. (eds). ISTE Ltd, London, and John Wiley & Sons, Hoboken.

Psola, J.H., Henke, M., Pandya, K. (2014). Characteristics of energy storage in smart grids. *9th International Conference on Ecological Vehicles and Renewable Energies, EVER 2014.*

Pye, J.D. (2008). *System modeling of the compact linear Fresnel reflector. Thesis*, University of New South Wales, Sydney.

Pyloudi, E., Papantoniou, S., Kolokotsa, D. (2015). Retrofitting an office building towards a net zero energy building. *Adv. Build. Energy Res.*, 9(1), 20–33.

Rabl, A. and Rialhe, A. (1992). Energy signature models for commercial buildings: Test with measured data and interpretation. *Energy Build.*, 19(2), 143–154.

Rahbari, O., Vafaeipour, M., Omar, N., Rosen, M.A., Hegazy, O., Timmermans, J.-M., Heibati, S., Van Den Bossche, P. (2017). An optimal versatile control approach for plug-in electric vehicles to integrate renewable energy sources and smart grids. *Energy*, 134, 1053–1067.

Rawlings, J., Coker, P., Doak, J., Burfoot, B. (2014). Do smart grids offer a new incentive for SME carbon reduction? *Sustain. Cities Soc.*, 10, 245–250.

Rivarolo, M., Greco, A., Massardo, A.F. (2013). Thermo-economic optimization of the impact of renewable generators on poly-generation smart-grids including hot thermal storage. *Energy Convers. Manag.*, 65, 75–83.

Rizos, I. (2007). Next generation energy simulation tools: Coupling 3D sketching with energy simulation tools. PhD thesis, Energy System Research Unit, University of Strathclyde, Glasgow.

Roberts, B.P. (2008). Sodium-sulfur (NaS) batteries for utility energy storage applications. *IEEE Power and Energy Society 2008 General Meeting: Conversion and Delivery of Electrical Energy in the 21st Century, PES*, 1–2.

Roberts, B.P. and Sandberg, C. (2011). The role of energy storage in development of smart grids. *Proceedings of the IEEE*, 99(6), 1139–1144.

Rostampour, V. and Keviczky, T. (2017). Energy management for building climate comfort in uncertain smart thermal grids with aquifer thermal energy storage. *IFAC-PapersOnLine*, 50(1), 13156–13163.

Rousse, D.R., Ben Salah, N., Lassue, S. (2009). An overview of phase change materials and their implication on power demand. *2009 IEEE Electrical Power and Energy Conference, EPEC.*

Saffari, M., de Gracia, A., Fernández, C., Belusko, M., Boer, D., Cabeza, L.F. (2018). Optimized demand side management (DSM) of peak electricity demand by coupling low temperature thermal energy storage (TES) and solar PV. *Appl. Energy*, 211, 604–616.

Santamouris, M. and Kolokotsa, D. (2013). Passive cooling dissipation techniques for buildings and other structures: The state of the art. *Energy and Buildings*, 57, 74–94.

Scheff, P.A., Paulius, V.K., Huang, S.W., Conroy, L.M. (2000). Indoor air quality in a middle school, Part i: Use of CO2 as a tracer for effective ventilation. *Appl. Occup. Environ. Hyg.*, 15(11), 824–834.

Seppänen, O.A. (1999). Association of ventilation rates and CO_2 concentrations with health and other responses in commercial and institutional buildings. *Indoor Air*, 9(4), 226–252.

Shukla, A.K., Venugopalan, S., Hariprakash, B. (2001). Nickel-based rechargeable batteries. *J. Power Sources*, 100(1–2), 125–148.

Siano, P. (2014). Demand response and smart grids: A survey. *Renew. Sustain. Energy Rev.*, 30, 461–478.

Singo, T.A., Martinez, A., Saadate, S. (2008). Using ultracapacitors to optimize energy storage in a photovoltaic system. *SPEEDAM 2008 – International Symposium on Power Electronics, Electrical Drives, Automation and Motion*, 229–234.

Smith, S.C. and Sen, P.K. (2008). Ultracapacitors and energy storage: Applications in electrical power system. *40th North American Power Symposium, NAPS2008*, 1–6.

Soares, J., Silva, M., Sousa, T., Vale, Z., Morais, H. (2012). Distributed energy resource short-term scheduling using Signaled Particle Swarm Optimization. *Energy*, 42(1), 466–476.

Steinmann, W.D. and Eck, M. (2006). Buffer storage for direct steam generation. *Sol. Energy*, 80(10), 1277–1282.

STS–Med (2018). *Concentrated Solar Thermal Energy Systems Handbook*. STS-Med, Palermo.

Taibi, E., Fernández del Valle, C., Howells, M. (2018). Strategies for solar and wind integration by leveraging flexibility from electric vehicles: The Barbados case study. *Energy*, 164, 65–78.

Thatte, A.A. and Xie, L. (2012). Towards a unified operational value index of energy storage in smart grid environment. *IEEE Trans. Smart Grid*, 3(3), 1418–1426.

Theophilou, M.K. and Serghides, D. (2015). Estimating the characteristics of the Urban Heat Island Effect in Nicosia, Cyprus, using multiyear urban and rural climatic data and analysis. *Energy Build.*, 108, 137–144.

Thomsen, K.E., Rose, J., Aggerholm, S.O. (2011). *Principles For Nearly Zero-energy Buildings*. BPIE.

Torvinen, H. and Ulkuniemi, O. (2016). End-user engagement within innovative public procurement practices. A case study on public-private partnership procurement. *Ind. Manag.*, 58, 58–68.

Tselepidaki, I., Santamouris, M., Moustris, C., Poulopoulou, G. (1992). Analysis of the summer discomfort index in Athens, Greece, for cooling purposes. *Energy Build.*, 18(1), 51–56.

Tukey, J.W. (1949). One degree of freedom for non-additivity. *Biometrics*, 5, 232–242.

Uddin, K., Jackson, T., Widanage, W.D., Chouchelamane, G., Jennings, P.A., Marco, J. (2017). On the possibility of extending the lifetime of lithium-ion batteries through optimal V2G facilitated by an integrated vehicle and smart-grid system. *Energy*, 133, 710–722.

Utkarsh, K., Srinivasan, D., Trivedi, A., Zhang, W., Reindl, T. (2019). Distributed model-predictive real-time optimal operation of a network of smart microgrids. *IEEE Trans. Smart Grid*, 10(3), 2833–2845.

Valkering, P. (2013). How to engage end-users in smart energy behaviour? *3rd European Energy Conference*, E2C, APA.

Vasile, V., Petran, H., Dima, A., Petcu, C. (2016). Indoor air quality – A key element of the energy performance of the buildings. *Energy Procedia*, 96, 277–284.

Wilson, C., Hargreaves, T., Hauxwell-Baldwin, R. (2015). Smart homes and their users: A systematic analysis and key challenges. *Pers Ubiquit Comput*, 19, 463–476 [Online]. Available at: https://doi.org/10.1007/s00779-014-0813-0.

Wolsink, M. (2012). The research agenda on social acceptance of distributed generation in smart grids: Renewable as common pool resources. *Renewable and Sustainable Energy Reviews*, 16(1), 822–835.

Wong, N.H. and Khoo, S.S. (2003). Thermal comfort in classrooms in the tropics. *Energy Build.*, 35(4), 337–351.

Xiong, Y., Wang, B., Cheng Chu, C., Gadh, R. (2018). Vehicle grid integration for demand response with mixture user model and decentralized optimization. *Appl. Energy*, 231, 481–493.

Xue, X., Wang, S., Sun, Y., Xiao, F. (2014). An interactive building power demand management strategy for facilitating smart grid optimization. *Appl. Energy*, 116, 297–310.

Yang, T.C. and Hsiung, P.A. (2018). Model predictive optimization for energy storage-based smart grids. *Communications in Computer and Information Science*, 805, 153–165.

Yang, L., Yan, H., Lam, J.C. (2014). Thermal comfort and building energy consumption implications: A review. *Applied Energy*, 115, 164–173.

Zarza, E., Ajona, J.I., Hennecke, K. (1997). Project DISS, Development of a new Generation of Solar Thermal Power Stations. *8th International Symposium on Concentrated Solar Power and Chemical Energy Technologies*, 397–415.

Zhang, K.M., Thomas, R.J., Bohm, M., Miller, M. (2009). An integrated design approach for sustainable community development. *Proceedings of the 42nd Annual Hawaii International Conference on System Sciences, HICSS*, Waikoloa.

Zhang, J., Li, K.J., Wang, M. (2016). A bi-level program for the planning of an islanded microgrid including CAES. *IEEE Trans. Ind. Appl.*, 52(4), 2768–2777.

Zhao, H.X. and Magoulès, F. (2012). A review on the prediction of building energy consumption. *Renew. Sustain. Energy Rev.*, 16(6), 3586–3592.

Zhen, Y., Li, X., Zhang, Y., Zeng, L., Ou, Q., Yin, X. (2012). Transmission tower protection system based on Internet of Things in smart grid. *ICCSE 2012 – Proceedings of 2012 7th International Conference on Computer Science and Education*, 863–867.

List of Authors

Margarita Niki ASSIMAKOPOULOS
Physics Department
Section of Environmental Physics
and Meteorology
Group Building Environmental Studies
University of Athens
Greece

Nikolaos BARMPARESOS
Physics Department
Section of Environmental Physics
and Meteorology
Group Building Environmental Studies
University of Athens
Greece

Francesco CARDUCCI
Research for Innovation
AEA s.r.l.
Angeli di Rosora
Italy

Cristina CRISTALLI
Research for Innovation
AEA s.r.l.
Angeli di Rosora
Italy

Marina Kyprianou DRACOU
The Cyprus Institute
Nicosia
Cyprus

Niki GAITANI
Department of Architecture and
Technology
Faculty of Architecture and Design
Trondheim
Norway

Marco GENTILOZZI
Research for Innovation
AEA s.r.l.
Angeli di Rosora
Italy

Konstantinos GOBAKIS
Energy Management in the Built
Environment Research Lab
Environmental Engineering School
Technical University of Crete
Chania
Greece

Daniela ISIDORI
Research for Innovation
AEA s.r.l.
Angeli di Rosora
Italy

Nikos KAMPELIS
Energy Management in the Built
Environment Research Lab
Environmental Engineering School
Technical University of Crete
Chania
Greece

Theoni KARLESSI
Physics Department
Section of Environmental Physics
and Meteorology
Group Building Environmental Studies
University of Athens
Greece

Denia KOLOKOTSA
Energy Management in the Built
Environment Research Lab
Environmental Engineering School
Technical University of Crete
Chania
Greece

Marina LASKARI
Physics Department
Section of Environmental Physics
and Meteorology
Group Building Environmental Studies
University of Athens
Greece

Siew Eang LEE
Department of Building
School of Design and Environment
National University of Singapore
Singapore

Klemens LEUTGÖB
Managing Director
Engineering Office for Energy and
Environmental Technology
Vienna
Austria

Angeliki MAVRIGIANNAKI
Energy Management in the Built
Environment Research Lab
Environmental Engineering School
Technical University of Crete
Chania
Greece

Fabio Maria MONTAGNINO
IDEA s.r.l.
Termini Imerese
and
Consorzio ARCA
Viale delle Scienze
Palermo
Italy

Alaric MONTENON
The Cyprus Institute
Nicosia
Cyprus

Pietro MURATORE
IDEA s.r.l.
Termini Imerese
Italy

Anne Sigrid NORDBY
Architect
Asplan Viak AS
Sandvika
Norway

Alexandros PANTAZARAS
Physics Department
Section of Environmental Physics
and Meteorology
Group Building Environmental Studies
University of Athens
Greece

Filippo PAREDES
IDEA s.r.l.
Termini Imerese
and
Consorzio ARCA
Viale delle Scienze
Palermo
Italy

Andri PYRGOU
The Cyprus Institute
Nicosia
Cyprus

Mat SANTAMOURIS
School of Built Environment
University of New South Wales
Sydney
Australia

Martina SENZACQUA
Research for Innovation
AEA s.r.l.
Angeli di Rosora
Italy

Laura STANDARDI
Research for Innovation
AEA s.r.l.
Angeli di Rosora
Italy

Vagias VAGIAS
Energy Management in the Built
Environment Research Lab
Environmental Engineering School
Technical University of Crete
Chania
Greece

Luca VENEZIA
IDEA s.r.l.
Termini Imerese
Italy

Index

A, B

Accredited Professional (AP), 24
advanced metering infrastructure (AMI), 199
artificial neural network (ANN), 115, 136–138, 141, 237
building
  -integrated photovoltaic (biPV), 88
  automation system (BAS), 149
  Controls Virtual Test Bed (BCVTB), 149
  energy management system (BEMS), 39, 87, 142
  information modeling (BIM), 35
  management system (BMS), 3, 52, 106

C, D

coefficient of performance (COP), 86, 96, 235, 252, 253, 257
cogeneration of heat and power (CHP), 3
compound parabolic collector (CPC), 246, 253
concentrated solar power (CSP), 59, 231, 232, 244, 252, 257
day-ahead (DA), 227
Degree Days
  Cooling (CDD), 92
  Heating (HDD), 92
demand response (DR), 2–6, 9, 11, 15, 30, 152, 153, 222, 227, 261
demand side management (DSM), 219
direct current (DC), 89
  high voltage (HVDC), 230
direct normal irradiance (DNI), 89, 100
direct steam generation (DSG), 247
distributed energy resources (DER), 236
domestic hot water (DHW), 202, 207

E

electric vehicles (EV), 223, 225–227, 234, 261
  plug-in (PEV), 227
Energy Efficiency
  Directive (EED), 199, 200
  Ratio (EER), 86, 96, 235
energy management system (EMS) (*see also* building), 115, 141, 142, 150, 235, 244
Energy Performance Buildings Directive (EPBD), 13, 17, 18

energy signature (ES), 55–57, 66, 67, 83
energy storage
  compressed air (CAES), 228–230
  superconducting magnetic (SMES), 225
  systems
    electrical (EES), 222–224, 237, 241
    hybrid (HESS), 225, 227
    thermal (TES), 229, 231, 237, 239, 240, 249, 251, 252, 260–262
      aquifer (ATES), 233
      molten salt (MS-TES), 249, 253
extract, transform, load (ETL), 63, 65, 68

F, G

fan coil units (FCU), 86, 87, 105, 117
fuel cell (FC), 3, 222, 225
Functional Mock-up Units (FMU), 149
genetic algorithm (GA), 9, 223
grid-to-vehicle (G2V), 225
ground source heat pumps (GSHP), 237–240, 262

H

heat recovery units (HRU), 86–88
heat transfer fluid (HTF), 89, 232, 246–250, 257, 260
heating, ventilation and air conditioning (HVAC), 3, 15, 59, 85, 89–92, 96, 97, 100, 104, 113, 117, 144, 145, 149, 154, 157, 159, 162, 167, 169, 207, 235, 239, 240
human machine interface (HMI), 236
hydro-pumped systems (HPS), 228–230

I, K

incident angle modifier (IAM), 90
information and computer technology (ICT), 3, 7, 15, 141
information technology (IT), 1, 7, 142, 206
integrated design (ID), 13–17, 19, 21–24, 26–28, 31–37, 45, 49
integrated energy design (IED), 13, 14, 16, 19, 35
Internet of Things (IoT), 1, 10
k-nearest neighbor (KNN), 105
key performance indicator (KPI), 223, 237

L

lifecycle
  analysis (LCA), 21
  cost (LCC), 21, 23, 24, 30, 32
    assessment (LCCA), 24, 29
linear Fresnel
  collectors (LFC), 59, 88, 89, 101, 253, 260
  reflectors (LFR), 59, 89–92, 94, 96, 100, 246, 248, 249, 252, 257, 260
loss of load probability (LOLP), 230

M, N, O

mean absolute percentage error (MAPE), 73, 79, 83, 84
Member States (MS), 17, 18, 199
molten salt (MS), 231, 249–253, 260
National University of Singapore (NUS), 103, 104, 115, 116
nonlinear autoregressive network with exogenous input (NARX), 137–139
organic Rankine cycle (ORC), 233, 246, 248, 251, 252, 257, 261

P

parabolic trough collectors (PTC), 253
particle swarm optimization (PSO), 9, 223
Pearson's coefficient (R), 71, 137, 190
percentage of people dissatisfied (PPD), 132–134, 156, 162, 175–179, 185, 186, 191, 193, 194, 198
Performance Measurement Protocols (PMP), 151
phase change material (PCM), 233, 234
photovoltaic (PV), 9, 31, 44, 45, 49, 51, 59, 71, 72, 85, 86, 88, 91, 100, 208, 226–230, 234, 236, 239, 240, 244
point of delivery (POD), 235, 241
predicted mean vote (PMV), 132–134, 156, 162, 175–177, 184–186, 191, 193–195, 198
programmable logic controller (PLC), 89, 236

R, S

relative humidity (RH), 86, 103–110, 112, 113, 118, 127, 132, 135, 137–140, 147, 150, 156, 160, 175, 176, 178, 191
renewable energy sources (RES), 2, 6, 7, 15, 208, 225, 226, 228, 230, 237
Research for Innovation (RforI), 154, 155, 158, 159, 163, 166, 167, 169, 170, 172–181, 183, 188, 189, 191, 195, 197, 198
root mean squared error (RMSE), 74, 79, 84
School of Design and Environment (SDE), 115–118, 132, 137, 141, 142, 150
Single Coil Twin Fan (SCTF) system, 105
small and medium enterprises (SME), 9
Smart Grid Energy Research Center (SMERC), 227
solar heating and cooling (SHC), 31, 90, 96, 97, 100
Statistical Package for Social Sciences (SPSS), 155

V, W, Z

vanadium redox flow batteries (VRFB), 224
vehicle-to-building (V2B), 225
vehicle-to-grid (V2G), 225, 226
wind turbine (WT), 51, 234
zero-energy building (ZEB), 1–3, 5, 13, 15, 17–19, 59, 115, 141, 149
  nearly (NZEB), 2, 13–15, 17–19, 42, 58, 59, 85, 101

Other titles from

iSTE

in

Science, Society and New Technologies

2021

AIT HADDOU Hassan, TOUBANOS Dimitri, VILLIEN Philippe
Ecological Transition and Research

BARDIOT Clarisse
Performing Arts and Digital Humanities: From Traces to Data
(Traces Set – Volume 5)

BENSRHAIR Abdelaziz, BAPIN Thierry
From AI to Autonomous and Connected Vehicles: Advanced Driver-Assistance Systems (ADAS)
(Digital Science Set – Volume 2)

CARDON Alain
Information Organization of The Universe and Living Things: Generation of Space, Quantum and Molecular Elements, Coactive Generation of Living Organisms and Multiagent Model (Digital Science Set – Volume 3)

DOUAY Nicolas, MINJA Michael
Urban Planning for Transitions

GALINON-MÉLÉNEC Béatrice
The Trace Odyssey 1: A Journey Beyond Appearances
(Traces Set – Volume 4)

HENRY Antoine
Platform and Collective Intelligence: Digital Ecosystem of Organizations

LE LAY Stéphane, SAVIGNAC Emmanuelle, LÉNEL Pierre, FRANCES Jean
The Gamification of Society
(Research, Innovative Theories and Methods in SSH Set – Volume 2)

RADI Bouchaïb, EL HAMI Abdelkhalak
Optimizations and Programming: Linear, Non-linear, Dynamic, Stochastic and Applications with Matlab
(Digital Science Set – Volume 1)

2020

BARNOUIN Jacques
The World's Construction Mechanism: Trajectories, Imbalances and the Future of Societies
(Interdisciplinarity between Biological Sciences and Social Sciences Set – Volume 4)

ÇAĞLAR Nur, CURULLI Irene G., SIPAHIOĞLU Işıl Ruhi, MAVROMATIDIS Lazaros
Thresholds in Architectural Education (Engineering, Energy and Architecture Set – Volume 7)

DUBOIS Michel J.F.
Humans in the Making: In the Beginning was Technique
(Social Interdisciplinarity Set – Volume 4)

ETCHEVERRIA Olivier
The Restaurant, A Geographical Approach: From Invention to Gourmet Tourist Destinations
(Tourism and Mobility Systems Set – Volume 3)

GREFE GWENAËLLE, PEYRAT-GUILLARD DOMINIQUE
Shapes of Tourism Employment: HRM in the Worlds of Hotels and Air Transport (Tourism and Mobility Systems Set – Volume 4)

JEANNERET Yves
The Trace Factory
(Traces Set – Volume 3)

KATSAFADOS Petros, MAVROMATIDIS Elias, SPYROU Christos
Numerical Weather Prediction and Data Assimilation (Engineering, Energy and Architecture Set – Volume 6)

KOLOKOTSA Denia, KAMPELIS Nikos
Smart Buildings, Smart Communities and Demand Response (Engineering, Energy and Architecture Set – Volume 8)

MARTI Caroline
Cultural Mediations of Brands: Unadvertization and Quest for Authority (Communication Approaches to Commercial Mediation Set – Volume 1)

MAVROMATIDIS Lazaros E.
Climatic Heterotopias as Spaces of Inclusion: Sew Up the Urban Fabric (Research in Architectural Education Set – Volume 1)

MOURATIDOU Eleni
Re-presentation Policies of the Fashion Industry: Discourse, Apparatus and Power (Communication Approaches to Commercial Mediation Set – Volume 2)

SCHMITT Daniel, THÉBAULT Marine, BURCZYKOWSKI Ludovic
Image Beyond the Screen: Projection Mapping

VIOLIER Philippe, with the collaboration of TAUNAY Benjamin
The Tourist Places of the World
(Tourism and Mobility Systems Set – Volume 2)

2019

BRIANÇON Muriel
The Meaning of Otherness in Education: Stakes, Forms, Process, Thoughts and Transfers
(Education Set – Volume 3)

DESCHAMPS Jacqueline
Mediation: A Concept for Information and Communication Sciences
(Concepts to Conceive 21st Century Society Set – Volume 1)

DOUSSET Laurent, PARK Sejin, GUILLE-ESCURET Georges
Kinship, Ecology and History: Renewal of Conjunctures
(Interdisciplinarity between Biological Sciences and Social Sciences Set – Volume 3)

DUPONT Olivier
Power
(Concepts to Conceive 21st Century Society Set – Volume 2)

FERRARATO Coline
Prospective Philosophy of Software: A Simondonian Study

GUAAYBESS Tourya
The Media in Arab Countries: From Development Theories to Cooperation Policies

HAGÈGE Hélène
Education for Responsibility
(Education Set – Volume 4)

LARDELLIER Pascal
The Ritual Institution of Society
(Traces Set – Volume 2)

LARROCHE Valérie
The Dispositif
(Concepts to Conceive 21st Century Society Set – Volume 3)

LATERRASSE Jean
Transport and Town Planning: The City in Search of Sustainable Development

LENOIR Virgil Cristian
Ethically Structured Processes
(Innovation and Responsibility Set – Volume 4)

LOPEZ Fanny, PELLEGRINO Margot, COUTARD Olivier
Local Energy Autonomy: Spaces, Scales, Politics
(Urban Engineering Set – Volume 1)

METZGER Jean-Paul
Discourse: A Concept for Information and Communication Sciences
(Concepts to Conceive 21st Century Society Set – Volume 4)

MICHA Irini, VAIOU Dina
Alternative Takes to the City
(Engineering, Energy and Architecture Set – Volume 5)

PÉLISSIER Chrysta
Learner Support in Online Learning Environments

PIETTE Albert
Theoretical Anthropology or How to Observe a Human Being
(Research, Innovative Theories and Methods in SSH Set – Volume 1)

PIRIOU Jérôme
The Tourist Region: A Co-Construction of Tourism Stakeholders
(Tourism and Mobility Systems Set – Volume 1)

PUMAIN Denise
Geographical Modeling: Cities and Territories
(Modeling Methodologies in Social Sciences Set – Volume 2)

WALDECK Roger
Methods and Interdisciplinarity
(Modeling Methodologies in Social Sciences Set – Volume 1)

2018

BARTHES Angela, CHAMPOLLION Pierre, ALPE Yves
Evolutions of the Complex Relationship Between Education and Territories
(Education Set – Volume 1)

BÉRANGER Jérôme
The Algorithmic Code of Ethics: Ethics at the Bedside of the Digital Revolution
(Technological Prospects and Social Applications Set – Volume 2)

DUGUÉ Bernard
Time, Emergences and Communications
(Engineering, Energy and Architecture Set – Volume 4)

GEORGANTOPOULOU Christina G., GEORGANTOPOULOS George A.
Fluid Mechanics in Channel, Pipe and Aerodynamic Design Geometries 1
(Engineering, Energy and Architecture Set – Volume 2)

GEORGANTOPOULOU Christina G., GEORGANTOPOULOS George A.
Fluid Mechanics in Channel, Pipe and Aerodynamic Design Geometries 2
(Engineering, Energy and Architecture Set – Volume 3)

GUILLE-ESCURET Georges
Social Structures and Natural Systems: Is a Scientific Assemblage Workable?
(Social Interdisciplinarity Set – Volume 2)

LARINI Michel, BARTHES Angela
Quantitative and Statistical Data in Education: From Data Collection to Data Processing
(Education Set – Volume 2)

LELEU-MERVIEL Sylvie
Informational Tracking
(Traces Set – Volume 1)

SALGUES Bruno
Society 5.0: Industry of the Future, Technologies, Methods and Tools
(Technological Prospects and Social Applications Set – Volume 1)

TRESTINI Marc
Modeling of Next Generation Digital Learning Environments: Complex Systems Theory

2017

ANICHINI Giulia, CARRARO Flavia, GESLIN Philippe, GUILLE-ESCURET Georges
Technicity vs Scientificity – Complementarities and Rivalries (Interdisciplinarity between Biological Sciences and Social Sciences Set – Volume 2)

DUGUÉ Bernard
Information and the World Stage – From Philosophy to Science, the World of Forms and Communications (Engineering, Energy and Architecture Set – Volume 1)

GESLIN Philippe
Inside Anthropotechnology – User and Culture Centered Experience (Social Interdisciplinarity Set – Volume 1)

GORIA Stéphane
Methods and Tools for Creative Competitive Intelligence

KEMBELLEC Gérald, BROUDOUS EVELYNE
Reading and Writing Knowledge in Scientific Communities: Digital Humanities and Knowledge Construction

MAESSCHALCK Marc
Reflexive Governance for Research and Innovative Knowledge (Responsible Research and Innovation Set - Volume 6)

PARK Sejin, GUILLE-ESCURET Georges
Sociobiology vs Socioecology: Consequences of an Unraveling Debate (Interdisciplinarity between Biological Sciences and Social Sciences Set – Volume 1)

PELLÉ Sophie
Business, Innovation and Responsibility (Responsible Research and Innovation Set – Volume 7)

2016

BRONNER Gérald
Belief and Misbelief Asymmetry on the Internet

EL FALLAH SEGHROUCHNI Amal, ISHIKAWA Fuyuki, HÉRAULT Laurent, TOKUDA Hideyuki
Enablers for Smart Cities

GIANNI Robert
Responsibility and Freedom
(Responsible Research and Innovation Set – Volume 2)

GRUNWALD Armin
The Hermeneutic Side of Responsible Research and Innovation
(Responsible Research and Innovation Set – Volume 5)

LAGRAÑA Fernando
E-mail and Behavioral Changes: Uses and Misuses of Electronic Communications

LENOIR Virgil Cristian
Ethical Efficiency: Responsibility and Contingency
(Responsible Research and Innovation Set – Volume 1)

MAESSCHALCK Marc
Reflexive Governance for Research and Innovative Knowledge
(Responsible Research and Innovation Set – Volume 6)

PELLÉ Sophie, REBER Bernard
From Ethical Review to Responsible Research and Innovation
(Responsible Research and Innovation Set – Volume 3)

REBER Bernard
Precautionary Principle, Pluralism and Deliberation: Sciences and Ethics
(Responsible Research and Innovation Set – Volume 4)

VENTRE Daniel
Information Warfare – 2nd edition

Printed and bound by CPI Group (UK) Ltd, Croydon, CR0 4YY
15/09/2026
14962361-0003